1. Models and Elementary Mathematics

THE OPEN UNIVERSITY

Technology/Mathematics: A Second Level Course

**MODELLING
BY MATHEMATICS**
TM 281

Block 1
GRAPHS AND SYMBOLS

1. Models and Elementary Mathematics
2. Linear Models 1
3. Linear Models 2

Prepared by the Course Team

THE OPEN UNIVERSITY PRESS

THE MODELLING BY MATHEMATICS COURSE TEAM

David Blackburn   (Chairman)

Phil Ashby   (BBC)
Keith Attenborough   (Technology)
Gerald Copp   (Editor)
Peter Cox   (Student Computing Service)
Bob Davies   (Senior Counsellor)
Judy Ekins   (Mathematics)
Roger Harrison   (Institute of Educational Technology)
Don Hurtley   (Staff Tutor)
Maurice Inman   (Staff Tutor)
David Johnson   (Student Computing Service)
Roy Knight   (Mathematics)
Ernest Law   (Staff Tutor)
Owen Lawrence   (Staff Tutor)
Daniel Lunn   (Mathematics)
Patricia McCurry   (BBC)
Alistair Morgan   (Institute of Educational Technology)
Colin Robinson   (BBC)
John Sparkes   (Technology)
Bob Tunnicliffe   (Mathematics)
Mirabelle Walker   (Technology)
Geoffrey Wexler   (Technology)

The Open University Press,
Walton Hall, Milton Keynes
MK7 6AA

First published 1977. Reprinted with corrections 1981.

Copyright © 1977 The Open University.

Designed by the Media Development Group of the Open University.

Produced in Great Britain by
Technical Filmsetters Europe Limited, 76 Great Bridgewater Street, Manchester M1 5JY.

ISBN 0 335 06290 3

This text forms part of an Open University course. The complete list of units in the course appears at the end of this text.

For general availability of supporting material referred to in this text, please write to the Director of Marketing, The Open University, P.O. Box 81, Walton Hall, Milton Keynes, MK7 6AT.

Further information on Open University courses may be obtained from the Admissions Office, The Open University, P.O. Box 48, Walton Hall, Milton Keynes, MK7 6AB.

1.2

CONTENTS

AIMS

The aims of this unit are:

1 To introduce the idea of modelling as a purposeful representation of reality in a simplified form.

2 To show how models can be used, together with mathematics, to assist with prediction of future trends, with explanation of phenomena and with technological design.

3 To introduce bar charts and histograms as simplified methods of presenting data and to show that they are suitable for some purposes.

4 To revise and test the elementary mathematical skills that it is assumed you will have on entry to the course.

OBJECTIVES

When you have studied this unit you should be able to:

1 Distinguish between true and false statements concerning, or explain in your own words, the following terms:

bar chart	median
bias	modelling suppositions
continuous quantity	range
discrete quantity	rounding error
Gaussian or	square
normal curve	square root
histogram	standard deviation
horizontal axis	vertical axis
mean	

2 Distinguish between the characteristics and the purposes of given examples of models (SAQs 1, 2, 3).

3 State what is meant by a 'model' and give examples of what are, and what are not, models (SAQs 4, 5).

4 Describe the relationships between a given example of a real situation and mathematical models of particular aspects of it (SAQ 18).

5 Express a number to a given number of significant figures (SAQs 7, 8).

6 Construct a bar chart from a set of data. (SAQs 6, 8)

7 Choose between a bar chart and a histogram as a means of presenting data. (SAQ 14)

8 Construct a histogram to suit a particular purpose and use histograms for comparison or prediction purposes (SAQs 15, 16, 17, 19, 20).

9 Draw a sketch of a distribution which can be assumed to be Gaussian given its mean and standard deviation (SAQ 21).

10 Describe what you can infer about a distribution which approximates to the Gaussian form (SAQ 22).

11 Calculate (a) the mean and (b) the standard deviation of a set of data (SAQs 23, 24).

12 Determine the square and square root of a number (using a slide rule if you wish).

STUDY GUIDE

Your work for this first study week consists of reading this correspondence text, Unit 1 *Modelling and Elementary Mathematics*; watching the first television programme, TV1 *Piecing together a model*; completing the assignment work for Unit 1; some work with the slide rule, which will involve you in listening to the audio-disc, Disc 1, and studying your *Slide Rule Book*; and the optional diagnostic exercises with the computer.

Details of the work associated with this unit will be found in the items of the supplementary material, mailed to you with this unit.

This unit assumes no prior mathematical knowledge on your part except an ability to add, subtract, multiply and divide both positive and negative numbers, including decimals and fractions. These topics are briefly revised in Section 3 and if you can answer the self-assessment questions SAQs 9, 10 and 11 you will have fulfilled the requirements for the unit.

You should work through the correspondence text, section by section, answering the self-assessment questions as you come to them. The answers to all self-assessment questions are given at the end of the unit.

You are advised to have read the correspondence text before you watch the first television programme.

After studying the unit you can use the objectives and related self-assessment questions, to check what you are expected to have accomplished. The summary of the Unit is intended to help you when you come to revise.

1   THE IDEA OF MODELLING

Modelling is a key activity in all branches of science—including social science—in technology and in most applications of mathematics. It is a term you may well be unfamiliar with and so the first few pages of the course are devoted to explaining what the word means and how the course team will be using it.

This first unit is also concerned with ensuring that you have the elementary mathematical skills that you will need if you are to profit from the course. You are expected to be able to add, subtract, multiply and divide, so that after working with the concept of modelling for a while, you will be required to test yourself with some mathematical exercises.

The course will then continue to develop the two themes of modelling and mathematics, so that when you have completed it you will be in a good position to enter on a number of second-level Open University courses. You will find that even though the course has been produced jointly by the faculties of Mathematics and Technology it does not concentrate on the application of mathematics to technological issues. This is because modelling and mathematics are quite general activities. As a consequence, the course team hopes you will find that you can use the skills and understanding you acquire from this course in many fields, not just in technology.

1.1   Definition of a model

Models can take many forms, from the small-scale replicas of aircraft that are used in tests to see how a real, full-sized aircraft would behave in flight, to abstract diagrams representing a country's economy. All of them represent some part of reality in some simplified way, and they are all created for a particular purpose—often to do with guiding or predicting future behaviour patterns. Let me begin with a definition of the word 'model'.

A model is a simplified representation of some aspect of reality created for a specific purpose or purposes.

Some of you might want to ask what I mean by 'simplified' or 'reality' or 'representation', but you will see what they all mean in due course without too much difficulty, since I am not using them in any special sense.

I think the most important point to understand is that all models are made for some purpose, although the purpose may not be obvious, or it may be so obvious that you might have overlooked it. Throughout most of this course, one of the purposes of the models that you will be considering will be to make mathematical analysis possible. The mathematical analysis will have a purpose too, such as predicting population trends, unemployment, the weather, or how big a boiler your house needs to keep it warm. Because you have the freedom to create models in almost any way you like, if they further your stated purpose, successful modelling is an activity requiring skill, understanding and imagination, and since there are so many possible useful models, I think it will be helpful if I describe some commonly used examples.

1.2 Some common examples of models

A scale model, like the model aircraft mentioned earlier, is a very simple realistic kind of model. Figure 1 shows another example of a scale model; this time of a ship. It is being tested in a long tank in Teddington to see how much force is needed to overcome water resistance as it is drawn through the water at different speeds. It can similarly be tested for its stability in choppy or rough conditions. If such a model reveals a dangerous instability in a new design, or perhaps a tendency to plunge into large waves rather than ride them, then it may prevent a future disaster.

Figure 1   A model ship undergoing tests in a special tank.

Three hundred years ago, the *Vasa*, a new warship, the pride of the Swedish Navy, capsized in calm water on its maiden voyage out of Stockholm. A few tests, if they had then been possible, would have revealed the design fault.

The purpose of the model being tested at Teddington is to do with the technology of good design, though of course, scale models also allow you to imagine the appearance of the final 'reality'. The model is not concerned

Figure 2   The Swedish warship Vasa which capsized on its maiden voyage from Stockholm in the seventeenth century.

with internal layout, or radio control, or radar, or comfort—except any possible discomfort caused by the ship's instability—and any attempt to use the model inappropriately is likely to be misleading. It is not, in fact, a simple matter to ensure that the kind of testing done in the tank at Teddington is not misleading, but this is where skill and understanding come in.

These days, scale models of many kinds are used. Figure 3 shows the model of an estuary used to study the effects on silting of a proposed civil engineering construction.

Figure 3 *A model of Morecambe Bay. A tide generator is at the rear of the model.*

SAQ 1

SAQ 1

List some of the purposes for which you might construct a scale model of a proposed house. Both the internal arrangement of the rooms and the external features are included in the model.

All maps are models and they illustrate very well some of the freedoms you have when you create models. If you want to show the spatial interrelationships of countries and towns and oceans all over the world on a flat sheet of paper, how are you to do it? The world itself is not flat, but in order to represent it in a book, or show all of it at once, you have to spread it out on a flat surface.

Figure 4 shows three common ways of attempting to do this. Mercator's projection converts the longitude lines (the meridians), which converge to a point at the poles, into parallel north–south lines. The latitude lines, which are equally spaced rings on the surface of the globe, again become parallel lines and are no longer equally spaced on Mercator's projection. The result is that small areas (e.g. Great Britain) are only slightly distorted in *shape*, though their sizes relative to areas near the equator or near the pole are quite incorrectly represented. Greenland is shown as about the same size as South America, yet it is really only one-ninth its size. The most important property of this projection is that navigational directions can be drawn as straight lines on it. For example, if you were to fly due south west from London you could easily find, using Mercator's projection, where you would get to. All you need to do is draw a line on the map in a south-westerly direction from London, as shown by the red line in Figure 4(a). This line strikes South America at Paramaribo in Surinam, which is where you would actually arrive if you set a south-westerly course from London on an actual flight (or sea journey) over the

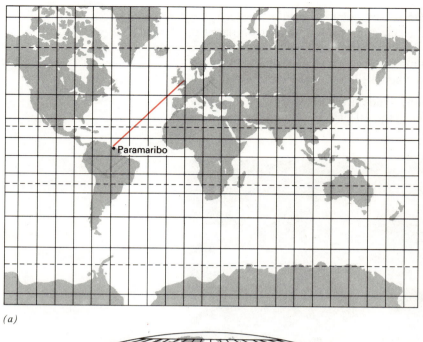

(a)

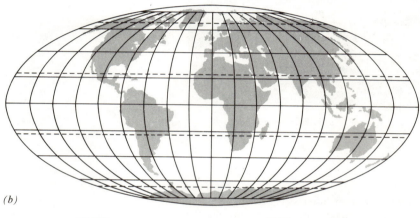

(b)

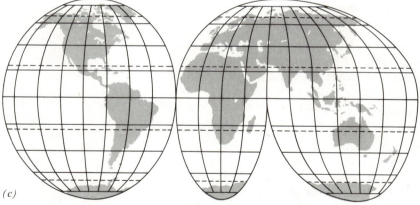

(c)

Figure 4   Three different projections of the map of the world. (a) Mercator's projection;
(b) Mollweide's projection; (c) Mollweide's interrupted projection.

curved surface of the Earth. On any other projection, courses on a fixed
bearing appear as curves. So Mercator's projection, despite its distortion
of size, is used for navigational purposes.

SAQ 2

(a) Does Mercator's projection represent the *shapes* of large areas like
     Canada correctly? If not, why not?

(b) Can you calculate the distance from London to Paramaribo by
     measuring the length of the line and then using a scale giving the
     number of kilometres per centimetre on the map?

Figures 4(b) and 4(c) show two other kinds of projections, called Mollweide's projection and Mollweide's interrupted projection. These are two models of the Earth whose purposes are to give a more realistic impression of the fact that the Earth is a sphere and to reduce the severe scale distortions of Mercator's projection. The interrupted projection, for example, represents the relative areas of the countries and continents fairly accurately, but their shapes are somewhat distorted. The shapes of the oceans are mercilessly sacrificed, however!

SAQ 3   SAQ 3

Figure 5 shows the standard map or representation of the London Underground system. State:

(a) What purpose it is intended to serve.

(b) What simplifications of the real situation have been introduced in order to achieve this purpose.

(c) An example of how you could be misled by it if you used it for the wrong purpose.

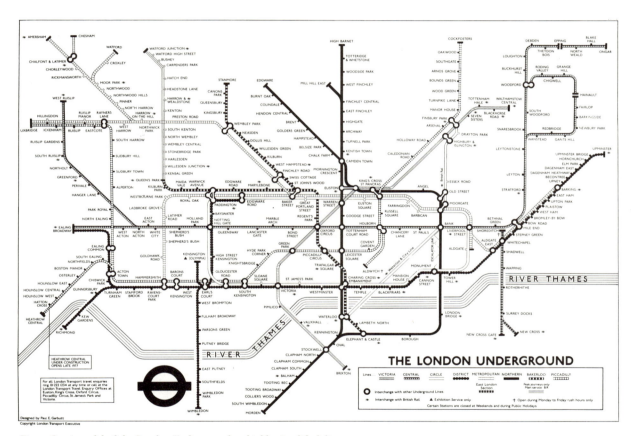

Figure 5   A model of the London Underground: a highly simplified diagram.

The lesson to learn from these illustrations is that in general, *almost any distortion or omission or approximation or assumption may be included in a model, so long as it furthers the intended purpose.* However, as I shall explain shortly, this is a bit too sweeping to cover scientific models, for their purpose generally is to reveal the truth about reality, so that deliberate distortions, for example, are not generally to be found in them.

For the moment I want you to concentrate on the idea of *simplification for a purpose.* The following example illustrates how a little ingenuity and imagination, based on understanding, can produce a simple model which uses elementary mathematics very simply and effectively.

A mathematical model

Your problem is to organize a singles tennis tournament. Suppose you have twenty-five entries and you have to arrange the tournament, with byes for those who do not have opponents in the first round, so that the whole tournament will fit into one afternoon. You have to know how many matches are involved and whether one kind of arrangement will work better than another.

The first thing to do is work out how many matches a particular arrangement involves. You can do this, perhaps, by drawing out the tournament as in Figure 6(a) and then counting up the matches. From this chart you can obtain information about who plays whom, as well as the number of matches. If all you want to know for the moment is the number of matches that have to take place, you can use a much simpler 'model' of the whole tournament—once it occurs to you. *It is true of any knock-out tournament that every match eliminates one player.* Since all but the winner must be eliminated, it follows that there are twenty-four matches if there are twenty-five competitors.

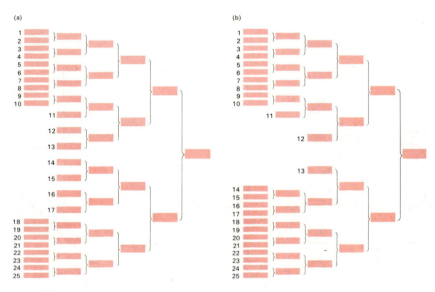

Figure 6   Alternative ways of arranging a knock-out tournament for twenty-five tennis players.

More generally, you can say that the number of matches will always be one fewer than the number of competitors. Later in the course you will find that it is normal to express such a statement as follows:

> If there are N competitors, where N stands for the actual number taking part, there will be $(N - 1)$ matches.

This simple way of looking at the problem makes it easy to calculate the number of matches and also shows that this number is not altered by how the tournament is arranged. Figure 6(b) shows an alternative tournament scheme, but the number of matches is the same as in the diagram of Figure 6(a).

This model, or simplified representation, of the tournament is called a *mathematical model* because it lends itself directly to the use of mathematics. The statement: 'In a knock-out tennis tournament each match eliminates one player', is the model. It is *so chosen* that it can be represented easily by mathematics. You can tell it is a model because it does not pretend to tell you everything about the tournament. It does not, for example, tell you who plays whom, as do the 'models' of Figure 6, and it is a mistake to attempt to use it for that purpose.

Approximate models

Note especially that the above model is not *approximate*; it is a *simplified* representation. It does not tell you the number of matches approximately, it tells you them exactly, but it does not tell you anything else. An *approximate* model might be: 'Every competitor plays an average of two matches, and each match involves two players'. From this you would conclude that with twenty-five competitors there would be fifty occasions in which everyone plays a match and since there are two players in each match there would be approximately twenty-five matches.

What is the correct number of matches?

24

This second statement about the tennis competition is therefore an approximate model.

The Underground map of Figure 5 is *simplified*, but it too is not approximate for its intended purpose. It tells you, without error, which stations interconnect two lines, but it does not tell you how to get to the BBC or the British Museum or the Palladium, because that is not its purpose.

So there is an important difference between a simplified model, which is accurate for its intended purpose, and an approximate model, which may, nevertheless, serve a different purpose better than the accurate model.

Graphs

To illustrate how approximate models can be helpful, consider the information about Open University applicants in Table 1.

This set of numbers can be represented by what is probably the commonest of all mathematical models—a graph. Figure 7 shows a graph of these data, in which the numbers of applicants are represented by lengths up the page and the years by lengths across the page. Each cross on the graph indicates one row of the information given in Table 1. This, of course, is a very simple form of modelling, and can be made as accurate as it is possible to measure lengths on the paper.

Table 1 Open University applicants 1971–76

year of entry	number of applicants
1971	43 444
1972	35 182
1973	32 046
1974	35 011
1975	52 537
1976	52 551

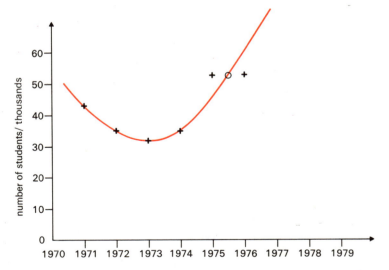

Figure 7 Numbers of Open University applicants from 1971 to 1976. The curve shows a possible trend.

The approximate model I was referring to above, is the continuous curve drawn through the points. The curve is drawn so that it is smooth, without the sudden kinks or corners it would have if it was drawn directly from one

15

cross on the graph to the next. It therefore indicates a *trend*. Any line drawn from one year to the next would seem to suggest that the number of applicants varies through a year. For example, half-way between 1975 and 1976 there is a circled point on the curve which you might suppose indicated that there were 52 544 people applying to start Open University courses in June 1975—which is nonsense; the university does not start courses in June.

So what does the curve represent?

It represents *approximately* what seems to be a *trend* in the numbers of applicants from one year to the next. The purpose of the model is to try to estimate the numbers of applications likely for 1977, 1978, and so on. It represents what past history indicates the future might hold in store. You do not need to draw a curve to conclude from the data that the number of applicants will not reach one million nor, on the other hand, that it will not drop to zero. By drawing the smooth curve as shown, in a direction indicated roughly by the points marked, you may conclude that for 1977 there will be 70 000 applicants, give or take a few thousand, and even more in 1978. This may turn out to be wrong, but unless there are reasons which suggest a significant change in the behaviour of potential students, using this approximate model is the simplest way of estimating future student numbers.

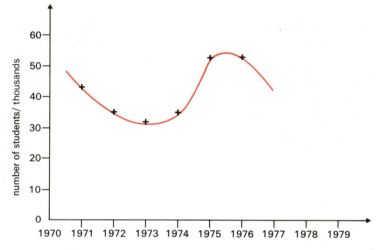

Figure 8   The same data as in Figure 7, but showing a different possible trend.

The curve—that is, the model—could, however, be chosen differently. Figure 8 shows a different curve through the same data which suggests a rather different kind of future; but either curve is better than having no idea at all of what the future holds.

SAQ 4
SAQ 4

Figure 9 shows data points for the population of Great Britain at several times this century. Estimate, by drawing an approximate curve through these points, what the population is likely to be in 1981 and 1991. What reasons, if any, have you for doubting the validity of your estimate?

Empirical and causal models

If you have no relevant data to use for estimating future trends, how can you make predictions? How, for example, could the Open University have predicted what the demand for its courses would be?

16

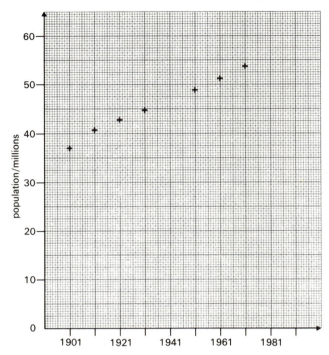

Figure 9   The population of Great Britain from 1901 to 1971.

One way would be to carry out a survey, to ask a representative group of the population whether they would take such courses. If one person in fifty said 'yes', you could estimate, very crudely, that in a population of fifty million there would be about one million potential students. Another way would be to describe what sort of person would take Open University courses and to estimate how many people in the country fit this model of the typical Open University student. Both methods involve the idea of modelling.

Where does the modelling process come into play in the first method—using a survey?

You have to decide on your representative group. Supposing you decide to ask 5000 people; how do you choose them? Do you ask people in the street? Do you make house-to-house surveys in a 'typical' small town? Do you classify all the social groups (e.g. working class, middle class, etc.) and deliberately choose 5000 people so that the balance of social groups is the same as for the country as a whole? Do you pick 5000 names at random from a telephone directory?

SAQ 5

SAQ 5

The above paragraph suggests four quite different ways of creating a model of the population (i.e. a representative group). For each suggestion, think of a reason why it might give you a misleading impression about the population as a whole. Remember what the purpose of the survey is. Try to think of a different reason for doubting each suggestion.

To give an example of the kind of answer I want from you, here is a reason why some of the suggestions could be misleading. The Open University does not take students under twenty-one years of age, so if the survey contains people under twenty-one who answer differently from those over twenty-one, the answers will be misleading.

This is not the place to answer the extremely difficult question about which representative 5000 to choose. It is one of the problems that social scientists grapple with. (Actually, the Open University commissioned a market research company to carry out a small survey in the south east of England. It uses its own methods which, if successful, keep it in business.) I have used this example only as an illustration of the importance, and the difficulty, of good modelling. Models of this kind which rely on measurement of numbers of chosen population may be called *empirical* models.

Let us now look at the other method of predicting Open University student numbers—describing a model of the typical Open University student. Such a model may be described as a *causal* model or a *theoretical* model.

There are several possible reasons for describing the characteristics of a typical Open University student, or for describing two or three different 'typical' Open University students. One purpose is to enable a course team to decide what prior knowledge it is sensible to assume that a student possesses. (You will come across the consequences of the course team's estimate of your lowest acceptable mathematical ability, shortly!). The purpose, in the present context, of describing typical Open University students is to estimate the number of *potential* students in the country, *without* carrying out a survey. If we can describe the sort of people who might take an Open University course (that is, describe some typical models of Open University students) we might be better able to estimate the percentage of the population that might become students. In particular, we might be able to list some characteristics that we can turn into numbers by going to a library or writing to some institution for information (e.g. the number of people who did not go to university). For example, one kind of model Open University student might be: 'Housewife, intelligent, gave up career to get married or have children, thinking about starting a career again, feels she is turning into a cabbage'. Such a model certainly restricts the range of possible applicants very considerably as compared with the whole population, although the proportion who feel they are turning into cabbages can be no more than a guess. You can draw up similar 'models' for other typical students.

> Try writing down a description of a broad class of students which you feel includes yourself.

This is a simplified representation of a part of reality for the purpose of estimating student numbers. It is therefore a model. It can be called a *causal* model because it identifies possible *causes* or *reasons* for being a student. It is not concerned directly with counting numbers as were the empirical models.

Distinctions between models, descriptions and data

The last example, like the previous ones, should be giving you the feeling that you understand what I mean by the process of modelling. It has a wide range of valid interpretations, as you have seen. However, it is not helpful to use the word 'model' to stand for all kinds of description, so let me explain how you can distinguish, broadly speaking, what the course team will not include as models. It is impossible to give hard and fast rules for there will always be grey areas, but some distinctions are worthwhile.

Descriptions must contain some element of active, often deliberate, simplification-for-a-purpose in order to be called models. For example, atoms are now known to be very complicated things, but for some purposes it is helpful to think of them as just tiny elastic particles. This is a simplified model; the full description, including neutrons, mesons, etc., is intended to be an accurate and complete description of reality and is not thought of as a model of reality. It represents reality as it is. The theory of gravitation was also intended to describe reality—not a simplification of it. Such attempts at fully accurate descriptions of reality are not regarded as models.

Numerical data are not thought of as models. I suppose you could argue that to say that the population of London is ten million is a simplified description-for-a-purpose, but the simplification is extreme and the purpose so vague and unspecified that to do so extends the meaning of modelling beyond what is sensible. Only when you represent that number by a point on a graph, or on a chart, or in some other way for the kind of purpose just discussed, is it helpful to call it modelling.

A careful interpretation of the definition of a model is a good guide to the meaning of the word, but in general, it is not profitable to argue at length about borderline cases because the word is not intended to have a very precise meaning.

Scientific, technical and predictive models

One final point, although modelling in one form or another is a normal part of many activities, it is primarily in fields such as technology that the idea of modelling has to be given special attention. We all tend to create images or models of the world around us to replace knowledge, for knowledge about the future can never be certain, though it is vitally important to us. We anticipate the way people will behave, or whether it will be safe to cross a road; we think of Americans as wealthy, or the English as hypocrites. Many of our actions are guided by images or models of reality of this kind, even though some of them are quite false. Much of scientific advance is concerned with increasing our knowledge of the world, and so with providing us with more reliable and accurate models with which to anticipate future events.

The aim of technology includes this aspect of modelling, but is also rather different, for it is concerned with *designing* some aspects of the man-made world. The problem then is to bring all available knowledge of science, of people, of structures, of the environment and so on, to bear on the design so that it will be successful.

In the past, technology has concentrated rather strongly on the scientific and mathematical aspects of engineering and has allowed market forces to influence people as to what they value sufficiently to buy. It has ignored any thoughts that the design might adversely affect the environment or the quality of people's lives, or make unrealistic demands on resources. Indeed, rather than get its designs right, industry or commerce has tended rather too often to resort to advertising to ensure good sales, even of inferior designs.

Be that as it may, the designer, in order to do his job successfully, has to relate a number of dissimilar aspects of the real world, as I have indicated, and he therefore needs simplified models of each aspect if he is not to become overwhelmed by the problem. This is why he creates these mental images of reality.

It is just as important that the designer does not oversimplify. The idea that rivers can absorb all the waste we discharge into them is now known to be over-simplified: 'a well-flushed sewer' is not a good model for a river on which to base the drainage system of a town or city. The idea that young families can adapt to modern high-rise accommodation is also over-simplified. No-one yet has a good model of human adaptability (not even of one's own ability to adapt to new circumstances) that can be relied on.

We know too, that the resources of fuels, materials and skilled manpower are limited. A model of future supply and demand is now needed. An effective designer of successful things in the man-made world needs simple reliable models of these aspects of his work, as well as models of scientific and technical knowledge, and he also needs an awareness that if he lacks sufficient knowledge to create a reliable model, the answer is not to ignore the issue, it is to provide breadth or flexibility in his design to cope with all relevant eventualities.

So there is a difference in the purpose of models in different activities. Science aims to give a true, accurate description of whatever is in need of explanation, and for many explanations simplified, approximate models are adequate. Technology is concerned with designing, making, controlling the man-.nade world. Many kinds of people are concerned with prediction, and this may only involve estimating a trend from a set of data without any particular understanding.

The test of a scientific model is whether it is true. The test of a technological one is whether it is successful. The test of any predictive model in any field—population, economics or whatever—is whether it turns out to be right.

In all three kinds of modelling, mathematical models play a key role.

1.3 Summary

1   A model is a simplified representation, created for a specific purpose (or purposes), of some aspect of reality.

2   Models can be used for a variety of purposes ranging from helping us to think clearly and constructively, to judging the appearance of things, predicting the future or designing new processes, etc.

3   Models do not necessarily have to be accurate representations in order to fulfil their stated purpose.

4   For predicting the future two kinds of modelling can be distinguished:

(a) One which isolates the subject of interest (e.g. a population), measures the property of interest (e.g. the birth rate, or Open University student application rate) over a period in the past, and then assumes that the trend will continue into the future. Modelling takes place in choosing the group of people to observe and in representing several isolated numbers as part of a continuing trend. These might be called empirical models.

(b) One in which the *causes* of the phenomena to be predicted are isolated. If the causes are understood, even in a simplified form, future trends might be predictable (e.g. population growth might be predictable from the tendency of parents to have, say, three children per family; or the Open University's student population might be predictable from understanding the reasons for study.)

2 MODELLING USING MATHEMATICS

Now that you are, I hope, becoming more familiar with the concept and purpose of modelling, and with the nature of various kinds of model, I want to describe the process of modelling using mathematics. I have called it *the modelling cycle*. This term simply refers to the idea that the modelling process is a two-way process or, more precisely, a circular process, as illustrated in Figure 10.

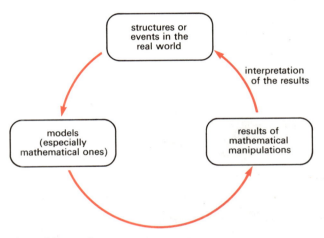

Figure 10   The modelling cycle.

The process begins with a real problem. It may be a problem of explanation or one of planning or one of design. In order to solve the problem, or to arrive at acceptable answers to the questions posed by the problem, you take three primary steps, each of which may have parts or small steps within it.

1   You devise a simplified model which represents what you believe to be the essential parts of the real situation, and which is in such a form that it can be expressed in mathematical form.

2   You then make some deductions using mathematics.

3   You then translate this solution back to reality: you have to interpret your mathematical results.

   If you are involved in Technology you will probably have designed something. If you have done your modelling well, your deductions should tell you something about the real situation which was not apparent in the first place and will have contributed to the solution of your problem.

   This return to reality is the test of the success of the other steps.

A scientist, say, performs some experiments on the behaviour of gases. He represents his observations with an atomic model (e.g. that atoms are tiny elastic particles). He expresses the behaviour of his model mathematically and calculates some new results. He translates these mathematical results concerning his model back into reality to predict results of new experiments. If all this has gone well his prediction will be confirmed by further observations.

One of the key strategies that technologists use in completing this modelling cycle is to make things 'in the image of' their mathematical or scientific models. It is obvious enough, once it is pointed out, that whereas animals have complex shapes and structures all intermingling and interacting in complex ways, machines are made of parts which interconnect in much simpler ways, which are made of specially prepared materials and are normally in shapes which have straight or circular edges, flat or circular sections, smooth surfaces, and so on. One reason for this is that the mathematics for calculating the properties of components made of homogeneous materials with straight or circular sections is much easier to handle.

Suppose your task is to design a form of transport. You require a method of moving heavy loads. You could try to copy the four-legged movement of animals, but simpler movements along straight lines or smooth curves can be analysed and described by much simpler mathematics, and this leads to the design of smooth rails, gentle curves, circular wheels, cylindrical pistons, etc.

The second reason for making things this way is that it is much easier to do so than to attempt to make living tissue!

Figure 11   The use of geometric shapes in technological structures.

Thus, part of the modelling cycle involves actually shaping reality to resemble the simple mathematical models you have chosen to help you solve your problem. Conversely, if this is to be your strategy, you do not choose mathematical models which you know you cannot copy in manufacture. The consequence of these procedures is, of course, that there is often a strong resemblance between man-made structures and the simple mathematical shapes used to design them. Rectangular tower office blocks, electricity pylons, girder bridges are examples of simple mathematical models being turned into reality (Figure 11). It is also a consequence of this procedure that mathematical models, *after* the structures have been built, are sometimes quite good descriptions of them.

All this leads to the idea that technology is applied mathematics. However, I think you can see now that mathematics, as well as science, human needs and wishes, economics, etc., are all brought into play together during the design process, in order to create an acceptable design. The resemblance between technological structures and mathematical models arises as much from choosing the mathematics which will describe what you can make, as from making what the mathematical models indicate.

Let me redraw the modelling cycle of Figure 10, taking into account these further thoughts: the result is shown in Figure 12. The cycle begins and ends with the real world. In science and economics, for example, the end product is perhaps an explanation or prediction: in technology, the end product is a decision about a course of action or perhaps a new or modified part of the world, shaped somewhat in the image of the mathematics used. Notice, however, that now the modelling cycle has a web of interactions superimposed on it. The models and the mathematics you use depend on what is to be explained or manufactured or organized in the world.

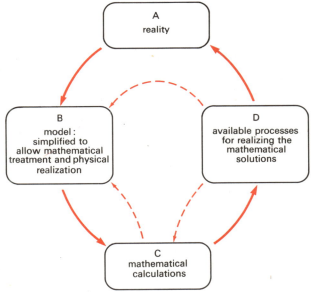

Figure 12   *The modelling cycle, the model takes into account the mathematics to be used as well as the way the end product is to be realized.*

Not all modelling by mathematics is for purposes of design or technology. If the purpose of the model is not to create something new, but rather to predict future events or to explain phenomena scientifically, for example, the emphasis in the model is much more towards accuracy. Without the need to keep an eye on what is practicable, one need not search so carefully for valid simplifications. Even so, modelling by mathematics is a cycle. It is necessary to interpret the mathematics.

To illustrate what I mean, let me recount a story which is sometimes told about a mathematician and an engineer. Each is asked to imagine he is placed two metres from a bottle of whisky and that they can each take one step towards it per second, but that each step must exactly halve the distance from it. They are then asked how long it would take to reach the bottle.

The mathematician's answer was: 'Never, there would always be a small distance separating me from the bottle'.

The engineer's answer was: 'About four seconds. That would bring me close enough for all practical purposes'.

You can see that each had accepted the model implied by the rather hypothetical question—that it makes sense to describe footsteps and distances from an object in this way. You would have to have things clarified a good deal to put the 'experiment' into practice (e.g. are we talking about the distance from toe or hand or mouth, or what?), but at any rate, both accepted the model implied by the question.

Next, both were able to do the mathematics, that is, divide two metres by two in successive steps.

In accepting the mathematical model, the mathematician had left the real world behind and chose not to return to it: he did not interpret his model. The engineer, through bitter experience, knew that this mathematical model, like most models he uses, is not a perfect fit with reality and remembered that the mathematical calculations were not worth pursuing further than the limits of precision of the model. For the mathematician in this story, the mathematical expression of the problem had become the reality of the situation.

For the scientist, the technologist, the social forecaster, the return to reality is an essential step that he must keep in mind.

2.1 The use of numerical data

The purpose of this section is to illustrate how numerical data can be represented and to show that, even in such a simple process as counting, the choices you make about how to describe reality can affect the numbers you use quite strongly and so affect your conclusions. The desire to reduce complex situations to numbers and mathematical concepts is very strong these days because of the remarkable achievements that the use of mathematics has brought about, but mathematical modelling is often oversimplified. The desire to obtain some result can lead to too great a simplification of the model.

The simplest of all ways of using mathematics is through counting. Population, production units and so on, become numbers simply by counting. (Another common way is by taking measurements, which I shall discuss further in Section 4.) Even in counting there can be problems. The concept of a population seems simple enough and hardly needs a description beyond a dictionary definition: 'the number of inhabitants of a country', but difficulties or uncertainties can arise. Inhabitants are presumably living, separate individuals and life begins at birth. Yet there are those who believe that the life of an individual begins at conception, not at birth, so should we argue that pregnant mothers should count as two? From a mathematical point of view there is no need to enter the argument, you simply state that life begins at birth and count the number of living separate individuals, or that life begins at conception and count pregnant women twice (or maybe three or four times if you can pick out those with twins or triplets). All you need is a clear statement to work with.

Perhaps you regard this example as a quibble, but what about 'the working population'. It is a phrase often used, but who is a member of the working population? For the purposes of some calculations, housewives who do not have a paid job are clearly not to be regarded as members of the working population—they are supported by their husbands. Children and old age pensioners are not included either; but what about those with part-time jobs; what about the self-employed? How much must you earn as a window cleaner, or, for that matter, by selling your own free range eggs, or Tupperware, or fresh honey, before you can be counted as a worker?

To convert the members of a class of people into a number you have to eliminate the grey areas round the edges of your classifications; who is a worker and who is not. This is one of the problems with mathematical modelling. Most of the time it does not occur to you to make these classifications, since for everyday life they are not needed. The fact that Tommy Smith doing his paper round before school blurs some neat classification is not really important. It is only important if you want to count those within a class exactly—for some political purpose, say.

This kind of distortion of reality (that is, forcing a decision when neither alternative is exclusively true) can be one of the penalties of modelling by mathematics. It is like trying to decide whether orange is to be classed as yellow or red. It is neither, but for some purposes you might want to put it on one side or other of a clear demarcation line.

SAQ 6 SAQ 6

Table 2 shows the number of people in each of six categories in a small town. Give possible reasons why the sum of these six numbers is more than the total population of the town.

Table 2   Distribution of occupations of the population of a small town

earning a living	4000
housewives	3000
old-age pensioners	300
unemployed	200
in full-time education	800
under school age	400
total population	7450

Let us suppose, however, that we have arrived at a set of categories that do not overlap and which have quite clear boundaries between one category and another. What is the best way of representing the numbers in these categories?

One way is simply to tabulate them, as in Table 3, which shows the state of affairs in an imaginary country in 1960.

Table 3   Distribution of occupations of the population of a country

working population (i.e. earning money)	25 575 000
unemployed (not in education)	390 000
in full-time education	10 130 000
housewives (not also working)	14 200 000
old-age pensioners (excluding wives of pensionable age)	7 880 000
under school age	4 200 000

It is difficult to get a good picture of the relative sizes of the various components of the population from a table of numbers, though if it is numerical accuracy you want, you cannot present the information in a better way.

Figure 13 shows the same numbers represented as 'bars' on a diagram. The diagram is called a *bar chart*, and in it the height of each bar represents the size of a different section of the population.

bar chart

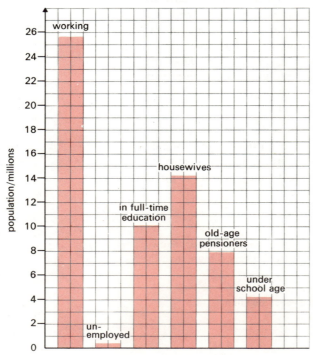

Figure 13 A bar chart of the distribution of occupations of the population of an imaginary country in 1960.

Figure 14 shows the same information for the population again, *together with* the sizes of the different sections of the population of the same country ten years later.

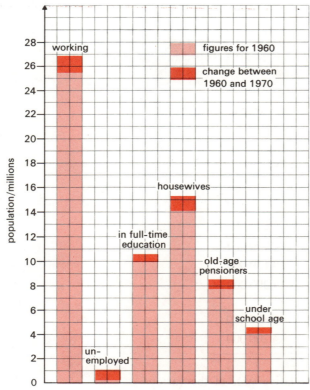

Figure 14 Bar charts showing how the distribution of occupations differed in 1970.

The construction of these charts, given the basic numerical data, involves converting the numbers into lengths, the sizes of which are proportional to the numbers.

This is done by drawing a line up the sheet of graph paper and marking on it equally spaced intervals each representing, in this case, one million inhabitants. This line is called the *vertical axis* of the bar chart. Each of the categories is then represented by a bar and reading off from the vertical axis at the height of the top of any bar should give the number of people in the category represented by that bar. For 1960, the longest bar (representing the working population) should be 25.6 divisions long (as read off on the vertical axis), and the shortest bar should be 0.4 divisions long. This means, of course, that the scaling of the vertical axis has to be chosen carefully so that all the bars, including the longest one, will fit into the space available.

vertical axis

Notice that I have said the longest bar should be 25.6 divisions long, although the working population divided by one million is 25.575.

> Why have I changed this number to 25.6 for the chart?

The answer is simply that it is not possible, with the naked eye, to see the difference between the two lengths 25.575 and 25.6. The difference between 25.5 and 25.6 is barely visible, so there is nothing to be gained by trying to draw a bar chart to any greater accuracy—since a bar chart's sole purpose is to present the given data in a clear visual form.

> Would you say that the bar chart would be misleading if the bar lengths were 26, 0.4, 10, 14, 8 and 4 divisions respectively?

I cannot think of any situation in which these approximate numbers, in a bar chart, would be misleading. Bar charts are inherently approximate representations of data, so it is no use trying to force them to be too accurate.

2.2 Accuracy and significant figures

One question which people are often asked when they move into a new house is how large their garden is. The sort of reply that they might give is: 'The front garden is quite small, but the back garden is a fair size. It's about twenty metres long'. This immediately gives the questioner an idea of the length of the back garden, which is quite adequate to satisfy his interest. He does not need or want to know that the garden is shown as 21.6 metres long on the plan of the house and garden supplied by the builder. In describing 21.6 metres as 20 metres, the owner of the garden may be said to have gone through the simple mathematical process of expressing 21.6 to *one significant figure*.

In order to express any given number to one significant figure, it is necessary to remember that, because of the way we write down numbers, the digit on the left represents the largest part of the number. For example, in 189 the 1 represents one hundred while the 8 only represents eight tens, or eighty, and the 9 represents nine ones, or nine.

Therefore, in expressing 21.6 to one significant figure it is the digit on the *left* you need to look at closely. In this case it is 2, representing two tens, or twenty. This would suggest that 21.6 can be expressed as 20 to one significant figure.

As a check, the digit to the right of the 2 must be examined. This is a 1, and 21 is closer to 20 than to 30, so 20 is the correct expression of 21.6 to one significant figure.

On the other hand, if the garden had been 28.3 metres long, then the digit on the left, representing two tens or twenty, is again most important, but this time a check on the digit to the right of the 2 shows this digit to be 8, and 28 is closer to 30 than to 20, so 30 is the correct expression of 28.3 to one significant figure.

While expressing the length of the garden to one significant figure may be sufficient to answer casual enquiries about the size of the garden, it may not be accurate enough when buying fencing for your new garden. You may need to measure the length of the garden to two significant figures. To express 21.6 to two significant figures you would look at the *two* digits on the left, in this case 2 and 1, representing twenty-one, and then check with the next digit to the right, which is 6. Since 21.6 is nearer to 22 than to 21, 22 is the correct expression of 21.6 to two significant figures.

For a garden 28.3 metres long, the two digits on the left are 2 and 8, representing twenty-eight, and a check with the next digit to the right shows it to be 3. Now 28.3 is nearer to 28 than to 29, so 28.3 may be expressed as 28, to two significant figures.

The difference between the original number and the answer obtained when it is expressed to a given number of significant figures is called the rounding error. Thus the *rounding error* in expressing 28.3 to two significant figures is $28.3 - 28.0 = 0.3$.

rounding error

While it is easy to see that 23.4 is 23 and 23.6 is 24, to two significant figures, one point that may have occurred to you is how to express the length of your garden to two significant figures if it is 23.5 metres long. Should it be expressed as 23 or 24, to two significant figures? This question only arises with the digit 5. By the convention used in this course, 23.5 would be expressed as 24.*

Summarizing, the digits 5, 6, 7, 8 and 9 are to be *rounded up*, while 0, 1, 2, 3 and 4 are to be *rounded down*. A couple of examples should make this process clearer.

Express 407 520 to three significant figures.

1 Look at the *three* digits on the left— 4, 0 and 7, representing four hundred and seven thousand.
2 Look at the next digit to the right, 5.
3 Round up because of the 5.
Answer   408 000.

Remember always to fill up with zeros. It is an easy trap to try to say that 407 520 is 408 to three significant figures, but this is clearly wrong— expressing to three significant figures cannot, and does not, convert four hundred and seven thousand-odd to four hundred and eight!

Express 4.991 to two significant figures.

1 Look at the *two* digits on the left— 4 and 9, representing 'four point nine'.
2 Look at the next digit to the right, 9.
3 Round up because of the 9.
Answer   5.0.

Notice that while 4.891 would have become 4.9 to two significant figures, 4.991 has become 5.0 because the second nine has caused the number to be nearer 5.0 than to 4.9. An interesting point is that to just *one* significant figure 4.991 would be 5 and not 5.0.

Can you see why this is so?

5.0 implies that both the 5 and the 0 are important—it implies *two* significant figures. The number 5, on the other hand, implies only *one* significant figure.

Expressing numbers less than one to a given number of significant figures follows a similar pattern, provided one important detail is remembered:

---

* This is not the only rounding convention in use in mathematics, but in this course, the course team will use, and ask you to use, this convention.

in 0.002 58 the leading zeros are only being used to hold the important numbers 2, 5 and 8 in their correct place and are not otherwise important themselves. The important digit on the left is the 2, and thus 0.002 58 is 0.003 expressed to one significant figure or 0.0026 expressed to two significant figures. With numbers less than one, the first step is always to start with the first digit on the left that is not a zero.

SAQ 7

SAQ 7

(a) Express 1250.6
  (i) to one significant figure;
  (ii) to two significant figures;
  (iii) to three significant figures;
  (iv) to four significant figures.

(b) Express 2.95 to two significant figures. What is the rounding error?

(c) Express 7390 to two significant figures. What is the rounding error?

(d) Express 0.0145
  (i) to one significant figure;
  (ii) to two significant figures.

  If you were to express 0.0145 to two significant figures and *then* to express the result to one significant figure, why would this be an unwise way of proceeding?

(e) Express 0.001 006 3 to three significant figures. What is the rounding error?

SAQ 8

SAQ 8

Draw a bar chart to represent the data in Table 4 showing the populations of the countries of Scandinavia in 1960. Before drawing the bar chart express the populations to a suitable number of significant figures and put the results in the third column in Table 4.

Table 4   The population of the countries of Scandinavia in 1960

country	population	approximate population
Sweden	7 495 000	
Norway	3 591 000	
Denmark	4 585 000	
Finland	4 446 000	

In taking measurements, an inevitable step in reading and recording the results is rounding up or down. If a pointer on a meter is between 18.5 and 18.6, but nearer 18.5 then you may wish to record this result to two significant figures (19) or to three significant figures (18.5) or you may wish to try to judge the position of the pointer between 18.5 and 18.6 and record the result as say 18.53 (which would be to four significant figures).

Sometimes the convention used in rounding numbers is not the one which will be used in this course. For instance, if you are asked your age, you normally give just the complete number of years—you would pro-bably *not* say you were 35 if you were 34 years and 7 months, even though 34 years and 7 months is nearer 35 than 34.

A different convention is also used in athletics, where races are timed to one thousandth of a second and are rounded up to the next higher one-hundredth of a second. A 100 metres run in 10.113 seconds would be recorded as 10.12 seconds rather than 10.11 seconds.

Weights of food may be rounded down to the next lower figure because the sale of under-weight goods is illegal, whereas no one minds a little extra for the same money.

Scientific measurements of length, weight, time, and so on, where accuracy is at a premium, are rounded in the way being used in the course.

Rounding down would exaggerate speeds in a sport where speed is at a premium.

2.3 Summary

1 The modelling cycle comprises:

   (a) representing some aspect of reality by a model which lends itself to mathematical treatment;

   (b) making some deductions using mathematics;

   (c) translating the results obtained back into a description of reality.

   The choice of model, the mathematics used and the final interpretation are normally interdependent.

2 When we make things which are based on science and mathematics they usually have much simpler structures and shapes than natural things. This is partly because mathematical models in terms of simple shapes and movements are easier to handle and fulfil the chosen purpose, and partly because we know how to make simple structures.

3 The description of reality in terms of distinct categories (e.g. workers and those who do not work) in order to count the number of members of each category can be a source of over-simplification and error.

4 The accuracy with which a number need be stated depends upon the purpose for which it is used. This accuracy can be expressed in terms of *significant figures*. (e.g. 533.32 can be expressed as 530 to two significant figures).

5 The process of changing a number from its most accurate form to one with a smaller, specified number of significant figures is referred to as 'expressing the number to a given number of significant figures'.

3 SOME ELEMENTARY MATHEMATICAL SKILLS

3.1 Mathematical requirements

This section is concerned solely with helping you to revise some important mathematical skills. The course does not set out to teach you the most basic mathematics, so I am going to assume that there are some skills you already possess. In fact, I assume that you can add and subtract, and can multiply and divide, that you can use decimal points, and that you know the symbol $+$ means add, the symbol $-$ means subtract, the symbol $\times$ means multiply, the symbol $\div$ means divide, and the symbol $=$ means 'is equal to'.

Further, it is assumed that you are not unfamiliar with fractions and negative numbers, but would benefit from some revision and explanation of them. Any further mathematical skills expected of you will be explained in this and subsequent units.

Thus, I assume you can carry out the following operations and obtain the answer given.

Addition $5.3 + 17.2 = 22.5$

Subtraction $172 - 53 = 119$

Multiplication $7 \times 21 = 147$

There is another sign meaning multiplication which you should know about. Brackets are used when several quantities are to be multiplied by the same number. Thus

$$7 \times 10 + 7 \times 5 - 7 \times 2$$
$$= 7(10 + 5 - 2)$$
$$= 7 \times 13$$
$$= 91$$

Many texts also use a full stop to indicate multiplication, but the course team will not do so, because it is so easily confused with the decimal point.

Division $44 \div 8 = 5.5$

Again, there are other signs to denote the operation of division. The oblique stroke or solidus

$$44/8 = 5.5$$

or underlining

$$\frac{44}{8} = 5.5$$

When several quantities are divided by the same number you can use brackets again. The following three examples are all equivalent mathematical statements.

$$\frac{40 + 4}{8} = 5.5$$

$$(40 + 4) \div 8 = 5.5$$

$$(40 + 4)/8 = 5.5$$

SAQ 9 (Revision)

Calculate:

(a) 467×28

(b) $28.5 - 14.2$

(c) $32.3 - 24.5$

(d) $176 \div 8$

(e) $241 \div 7$ (to three significant figures)

(f) $864 \div 27$

3.2 Negative numbers

I imagine you are familiar with the use of negative numbers to represent certain quantities in everyday life. Negative temperatures are used for temperatures below freezing point; negative bank balances indicate you are in the red; negative heights are used to indicate depths below sea level. They are used in science and technology for many other purposes for which you may not be so familiar, such as the use of negative velocities to indicate the speed of vehicles going backwards· or negative changes to indicate decreasing quantities. You will come across this idea frequently during the course. My main purpose in this section is to be sure that you will know how to handle negative numbers in various arithmetic manipulations without having to try to ascribe meanings to the negative numbers.

Test your understanding of the arithmetic manipulation of negative numbers by attempting SAQ 10. If you get stuck, the answers will help you.

SAQ 10 (Revision)

(a) $7 + (-3) =$

(b) $7 - (-4) =$

(c) $7 - (-14) =$

(d) $-7 - (-8) =$

(e) $-8 - 3 =$

(f) $-6 - (-5) =$

(g) $5 \times 3 =$

(h) $5 \times (-3) =$

(i) $(-5) \times 5 =$

(j) $(-5) \times (-4) =$

(k) $(-10) \div (-2) =$

(l) $9 \div (-3) =$

3.3 Fractions

Fractions are likely to occur when you measure things or when you divide. An office may be $3\frac{1}{4}$ metres by $4\frac{3}{4}$ metres. You may have £3 to divide equally between eight people, so that each person gets £3/8. Again, as with negative numbers, occasions will occur in this course when you will be required to manipulate fractions, irrespective of what they represent, and I expect you can already do this without assistance. To reassure you, I have set out some questions in SAQ 11, to which explanatory answers are provided at the end of the unit.

First, however, let me remind you of one or two facts about fractions.

1 $3\frac{1}{4}$ can be expressed as $\frac{13}{4}$.

2 In any fraction the top line is called the *numerator* and the bottom line is called the *denominator*. A fraction whose numerator is less than its denominator is called a proper fraction.

3 The fraction $\frac{12}{15}$ can be expressed as $\frac{4}{5}$. This is because you can multiply or divide both the numerator and the denominator (of a single fraction) by the *same* number without changing its value. In this case they have both been divided by the number 3.

4   Addition and subtraction of fractions is achieved by putting each fraction on the same denominator, called a *common denominator*, using the process described in (3) above.

5   Multiplying two fractions together is achieved by multiplying both the numerators together to form the numerator of the product, and then multiplying both the denominators together to form the denominator of the product.

6   One way of dividing by a fraction is to invert this fraction (i.e. interchange numerator and denominator) and then *multiply* by the inverted fraction.

SAQ 11 (Revision)

Evaluate

(a) $\frac{1}{2} + \frac{1}{4}$ (g) $\frac{1}{2} \times \frac{4}{5}$

(b) $\frac{1}{3} - \frac{1}{4}$ (h) $1\frac{1}{2} \times \frac{2}{3}$

(c) $1\frac{3}{4} + 2\frac{1}{8}$ (i) $2\frac{1}{3} \times 2\frac{1}{3}$

(d) $\frac{12}{16} + \frac{5}{20}$ (j) $\frac{4}{3} \div \frac{2}{3}$

(e) $\frac{13}{3} - \frac{7}{5}$ (k) $3\frac{1}{4} \div \frac{4}{7}$

(f) $5 - 1\frac{4}{5}$ (l) $1\frac{3}{4} \div 2\frac{1}{3}$

Suppose you are told that $\frac{3}{8}$ of the children in a school are girls and $\frac{5}{8}$ of the children are boys. If the school contains 720 children, how many of each sex are there?

The word 'of' in sentences about fractions always means multiplication, so the number of girls in the school is

$$\frac{3}{8} \times 720 = \frac{3 \times 720}{8} = 270$$

SAQ 12

In a school of 720, $\frac{5}{9}$ of the children are girls, $\frac{1}{8}$ wears spectacles, $\frac{1}{4}$ have blue eyes. How many of each category are there? How many boys, do you suppose, do not wear glasses *and* do not have blue eyes? (Think carefully before you try to answer the last question).

SAQ 13

A rather neat trick question you can ask your more mathematical friends, is the following. (The first part is the trick question, so do not spend long on it. The second part is the real exercise on fractions.)

(a) A man leaves his cattle to his three sons as follows: the eldest is to receive one-half of them; the second is to receive one-third; the third is to receive one-ninth.

When he died, the man left only seventeen cows. How do you suppose the three heirs divided up their legacy?

(b) Suppose the legacy was £17 000, not cattle. How much, to the nearest pound, should each son receive?

4 HISTOGRAMS

4.1 Discrete and continuous quantities

Bar charts, which you met in Section 2, are a useful way of representing some types of statistical data in a form that the eye can assimilate fairly easily. However, not all statistical data can be represented by means of a bar chart.

The distinction between the sort of data that can be represented by means of a bar chart and the sort that cannot is illustrated by Figure 15. Figure 15(a) shows the whole numbers or *integers* 0, 1, 2, 3, 4, 5 as a series of separate dots. Between 0 and 5, some numbers (such as 2.5 or 1.7) are not represented because there are no dots for them. The series of dots represents a series of *discrete* values—a series of separate and distinct values. Figure 15(b) shows a number line running from 0 to 5. Between 0 and 5 there is no number which is not represented by a point somewhere on the line: 2.5 is there, and so is 2.525 and 2.52525 and so on. No matter how many places of decimals you may give in your number, it is represented on the number line. This line represents *continuous* values—values which 'run into each other' so that it is not possible to find a number between 0 and 5 which is not represented on the line.

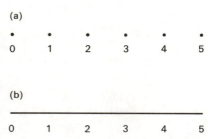

Figure 15 (a) The dots represent discrete values; (b) the line represents continuous values.

Bar charts are used to represent values which are discrete, as shown in the following example. A firm manufacturing skirts makes them according to a set of discrete sizes labelled 10, 12, 14, etc. Skirt size is a *discrete quantity*. This firm received the following sales figures one month (Table 5).

discrete quantity

This is the sort of information which can be represented on a bar chart, since the sizes of skirts fall into discrete values or categories. The bar chart is shown in Figure 16.

Table 5 Monthly sales figures for a skirt manufacturer

size of skirt	number sold
10	200
12	750
14	1100
16	1150
18	800

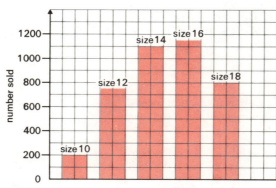

Figure 16 Bar chart showing the numbers of various sizes of skirts which have been sold.

Suppose now I consider the women who buy these skirts with the discrete categories of waist sizes. Do the women also have discrete categories of waist size?

As I am sure you are aware, there is a whole spread of values possible. In fact, women's waist size is a *continuous quantity*. Remember that I am talking about *actual* waist sizes. Even though waist sizes probably cannot be *measured* to more precision than about half a millimetre, that does not affect the fact that actual waist size is a continuous quantity.

continuous quantity

Each woman, when buying a skirt, puts herself into one of the discrete categories offered by the manufacturers and so the purchase of a skirt represents a process of grouping a whole range of values of a continuous quantity (women's waist measurements) into a few discrete categories (skirt sizes).

To represent a continuous quantity, such as waist size, we use a *histogram*. A histogram looks very much like a bar chart in some ways, but it also has important differences, as you will see. The process of preparing and drawing a histogram is something like the process whereby women group themselves into skirt sizes in one respect—the job of the person preparing the histogram is to subdivide some continuous quantity into groups or *classes* so that its distribution can be represented on the histogram. The way in which the groups are chosen will depend on the purpose in drawing the histogram.

histogram

Waist size is, of course, not the only example of a continuous quantity, nor are skirt sizes the only example of discrete categories. Other continuous quantities are heights or weights of people and other discrete categories are people's blood groups or glove sizes.

The point to remember is that a bar chart is a way of representing the distribution of items which fall into discrete categories and a histogram is a way of representing the distribution of a continuous quantity.

SAQ 14

SAQ 14

If you had collected data about the distribution of each of the quantities mentioned in (a) to (e) below, would you use a bar chart or a histogram to represent the data graphically?

(a) The size of shoe bought by the customers in a shoe shop in a given week.

(b) The weight of babies born in a hospital in a given period of time.

(c) Region-by-region sales of a make of car in the British Isles in one year.

(d) Annual incomes of the working population of London.

(e) The heights of a group of men applying to join the police force in one month.

4.2 Collecting data for a histogram

Suppose that a skirt manufacturer had been producing skirts more designed to appeal to 'the older women' and is thinking of branching out into skirts for 'teens and twenties'. He knows the distribution of sizes needed for the former category, of course, but feels that the distribution of sizes will not be the same for the new market. In fact, the manufacturer has a model, or mental picture, to the effect that younger women are slimmer, but does not know how much slimmer, nor even if the model is correct. So the manufacturer sets about instigating some market research on the subject. From this research it is hoped to gather data that will back up the model, and also quantify the word 'slimmer', or refute the model.

After discussion, a firm of market researchers are given the following brief:

1   The sample is to be 500 women in the age group 16–30.

2   The researchers are to take waist measurements only to the nearest centimetre.*

As shown in Section 2.2, this means that waist measurements of, say, 67.2 or 67.4 centimetres will be rounded down to the nearest whole number, 67, while those of, say, 67.5 or 67.8 centimetres will be rounded up to 68.

This is not the only way in which the manufacturer can find out about the distribution of sizes of skirts he needs to make. He can, instead, try to find out what his competitors in the 'teens and twenties' field manufacture, or make a guess and then refine his guess by examining sales figures and making more of some sizes and fewer of others. Since the manufacturer has chosen the method outlined above, he has to make some deliberate decisions.

Making decisions is usually a part of the modelling process. Sometimes it is necessary to choose to leave something out of the model for the sake of simplicity, sometimes it is necessary to suppose that something stays the same, even when it is fairly certain that in reality it does not stay quite the same, and so on. These decisions will be referred to as *modelling suppositions*, and you should look out for them whenever you meet a model in the course.

modelling suppositions

In this model, what are the modelling suppositions?

1   Women under 16 or over 30 may buy the 'teens and twenties' range of skirts, but the manufacturer has chosen to suppose they will not for the purposes of the model. (This is an example of a deliberate omission for the sake of simplicity.)

2   Far more than 500 women in this age range will no doubt buy the skirts, but the manufacturer has decided that a sample of 500, if chosen randomly from around the country, will enable him to make reasonable predictions about *all* the women who are part of the potential market, even though it may be that some of the 500 women will never buy the firm's skirts.

3   By asking for a waist measurement only to the nearest centimetre, the manufacturer is not asking for as great a precision of measurement as is possible, because the measurements can certainly be taken to the nearest millimetre and possibly even more precisely. But the *purpose* of the model is to find out how many sizes 10, 12, 14, etc., to produce and the manufacturer knows that measurements to the nearest millimetre are not needed to help to do this. The manufacturer also knows that women's waist sizes do not remain constant all the time to within a few millimetres, so there is no need to record waist sizes any more precisely than to the nearest centimetre.

4   The manufacturer has assumed that waist size is the only factor affecting the size of skirt women buy and that hip size, for example, does not matter.

These modelling suppositions are all forms of simplification and are all parts of the modelling process. It may be that different suppositions would lead to data that are more 'correct' when viewed in the light of subsequent sales figures and this is a risk the manufacturer has chosen to take. The expense and delay of starting with less simple criteria has to be weighed against the expense of making a mistake and being left with unsold stock.

This model is an empirical model because it relies on numerical data about a chosen group of people. In SAQ 5 you met some of the problems that can arise in choosing a representative group of people. Similar problems could arise here in choosing the 500 women in the 16–30 age group: should they be chosen by waiting at the door of a dress shop on the High Street; by a survey at some women's colleges; by a house-to-house survey in part of a town;...?

Can you think of reasons why each of these methods is unsatisfactory?

Each method is liable to introduce what is called *bias* into the survey. For example, most of the women at college are in the 18–21 age group, so

bias

results collected only from them would be biased towards this age-group and may not therefore be typical of the whole 16–30 range.

This should have drawn your attention to some of the pitfalls in carrying out a survey, but the manufacturer must presume that the market researchers he has engaged have methods of carrying out the survey so that the results are as unbiased as possible.

4.3 Drawing a histogram

Suppose that the manufacturer did go ahead on the basis outlined and received from the research team the figures shown in Table 6.

Table 6   Results of the survey on waist-sizes

waist measurement/ centimetre	number of women
57	2
58	0
59	5
60	7
61	7
62	8
63	11
64	17
65	26
66	31
67	35
68	38
69	46
70	50
71	43
72	38
73	36
74	29
75	24
76	14
77	11
78	0
79	9
80	7
81	0
82	5
83	1

These results are in an interesting form, in that waist measurement is a continuous quantity, but because of the way the manufacturer asked for the survey to be carried out it looks as though a discrete quantity is being recorded! In fact, the numbers in the 'waist measurement' column have been entered in what might be described as a 'shorthand' form, because a waist measurement recorded as 57 centimetres could have been anywhere in the range 56.5 to 57.5 centimetres, excluding 57.5 centimetres itself, but including any number just under 57.5.

What range of measurements will have been recorded as 71 centimetres?

All measurements from 70.5 to 71.5 centimetres, excluding 71.5 itself.

Because the manufacturer's purpose is to predict what proportion of the prospective market would buy the different sizes, and hence what proportions they should be manufactured in, the next step is to group the data.

Size 10 skirts have a 60 cm waist, size 12 a 65 cm waist, and so on, so the manufacturer has to make an extra modelling supposition about what sizes of skirt would be bought by women with different sizes of waist.* He could for instance decide that women with 58, 59, 60, 61 and 62 cm waists will buy size 10, and women with 63, 64, 65, 66, 67 cm waists will buy size 12, etc. Making this decision, the data are grouped as follows (Table 7).

Table 7 Results of the survey grouped according to waist-size

recorded waist sizes/cm	number of women
58, 59, 60, 61, 62	27
63, 64, 65, 66, 67	120
68, 69, 70, 71, 72	215
73, 74, 75, 76, 77	114
78, 79, 80, 81, 82	21

There is a problem with the two women with 57 cm waists and the one with an 83 cm waist. The former, according to the supposition, needs a size 8 and the latter a size 20, neither of which are manufactured. If he is to stick to his original brief, the manufacturer can only ignore these three measurements, leaving the manufacture of clothes for these women to firms which specialize in small or large sizes.

To represent the results of the survey graphically, the manufacturer would use a histogram because waist measurement is a continuous quantity. Figure 17 shows the histogram obtained.

Notice that numbers are written along the base line, called the *horizontal axis*, at regularly spaced intervals. This is possible because the quantity

horizontal axis

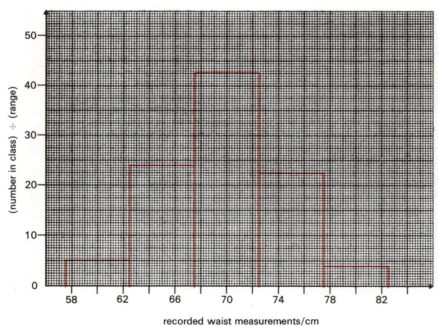

Figure 17   Histogram of the recorded waist measurements.

*cm *is the symbol for centimetre(s).*

being represented is continuous and is different from the bar chart representation, where items are in discrete categories and each column is separate and has a separate label. The scale is chosen so that the horizontal axis fits conveniently across the graph paper. In this case, 2 cm of waist size is represented by ten small intervals on the horizontal axis of the graph.

Notice also that the blocks rest on bases from 57.5 to 62.5 cm, from 62.5 cm to 67.5 cm, etc. This is because the measurements recorded as 58, 59, 60, 61 and 62 are in fact represent a range of measurements from 57.5 to almost 62.5, and so on. If one block finished at 62 and the next one started at 63, this would imply that those women with a waist measuring between 62 and 63 cm were not assigned to a category, which is not true.

The third point to notice is that the quantity plotted on the vertical axis is 'number in class divided by range'. The *range* is the difference between range the highest and lowest values in a class. Since each range in Figure 17 is 5, the heights of the blocks are all 'number in class divided by 5'. Once again a scale is chosen so that the vertical axis fits conveniently up the side of the graph paper—in this case 10 is represented by 20 small intervals. It may seem rather surprising to you that 'number in class' is not plotted directly on the vertical axis, but the choice of 'number in class divided by range' is conventionally used because by convention it is the *area* of a block in a histogram which represents the number of items in that block, and not the height. The reason for this use of area will become apparent when you do SAQ 17. By area I mean *scaled* area—I do *not* mean that you should put a ruler on Figure 17, measure the height and width of a block and multiply them together to obtain the area.

In the class 62.5–67.5 cm the width of the block, the range, in Figure 17 represents 5 and the height of the block represents 24. What is the area of the block and what does this represent?

The area is 24 × 5, so the area is 120. This is the number of women in the class 62.5–67.5 cm.

Once the manufacturer has a histogram which represents the results of the survey, he can use it to make predictions.

If the manufacturer commissioned a second survey with identical criteria, would an identical histogram to Figure 17 be obtained?

Almost certainly not; there is no reason to assume that any sample of 500 will be *exactly* the same as any other sample of 500, even if the two samples are chosen from the same total 'population' of women in the appropriate age range, but there *is* reason to assume that they would be very like each other.

The firm has to realize that the histogram does not tell it *exactly* what proportion of each size to manufacture; it only gives an indication. Assuming that the sample results give a fair prediction of the total group of women who would buy the skirts, the firm can go ahead with manufacture on this basis.

SAQ 15

SAQ 15

Suppose that the manufacturer had made a different decision about which women would buy which sizes of skirt. Suppose he had decided that a woman with a 62 cm waist would not squeeze herself into a size 10 skirt, but would instead buy a size 12, and similarly women with 67, 72, 77 and 82 cm waists would buy skirts which are a little too big.

Draw the histogram obtained if this had been the modelling supposition and compare it with Figure 17.

Would this decision have greatly altered the proportions of the different sizes he manufactured?

SAQ 16 SAQ 16

The manufacturer was about to expand his business into a country where the women are smaller than in Britain. He commissioned a survey and obtained the following data (Table 8).

Table 8   Results of a second survey on waist-sizes

waist measurement/cm	number of women
53	6
54	0
55	8
56	8
57	13
58	17
59	26
60	33
61	36
62	39
63	46
64	55
65	47
66	40
67	36
68	29
69	20
70	17
71	11
72	7
73	0
74	5
75	1
76	0

In the country in question skirts are sized A, B, C, etc., where size A has a waist of 54 cm, B has a waist of 57 cm, and so on.

Draw the histogram the manufacturer could have used in deciding what proportion of each size of skirt to manufacture.

SAQ 17 SAQ 17

Table 9 shows the age distribution of the population of the UK in 1971.

Table 9   Age distribution of the population of the UK in 1971

age	number	required data
under 5	4 522 000	
5–14	8 976 000	
15–19	3 895 000	
20–24	4 287 000	
25–39	10 148 000	
40–54	10 108 000	
55–59	3 345 000	
60–64	3 183 000	
65–69	2 690 000	
70 or over	4 514 000	

(a)   These data are to be represented graphically as a histogram. Use the blank third column in Table 9 to enter the data you will be plotting and then plot the histogram.

(Hints: (i) examine the sizes of the classes carefully before you start; (ii) remember that 5–14 means not under 5 and not equal to or over 15; (iii) '70 or over' is a vague term, making the choice of a range difficult. It is probably safe to assume that so few people live to over 80 that 80 is a fair upper limit to take for this class.)

(b)   Sketch out (but do not spend time plotting accurately) the so-called histogram obtained by plotting 'number in class' on the vertical axis and comment on why plotting 'number in class divided by range' gives a more useful representation of the data in Table 9.

SAQ 18 SAQ 18

Figure 12 of this unit gives a pictorial representation of the modelling cycle. In the skirt manufacturing model you have just read about, what can be represented in each of the boxes A–D?

4.4 Comparison using histograms

Histograms provide a useful way of comparing data. The figures in Table 10 show the heights of fifty men in each of two groups. Such data might be collected for research purposes, perhaps to test a theory that smaller parents tend to have smaller children, or that the presence or absence of some substance in a child's diet can affect his adult height, or that men from one part of the country tend to be smaller than men from another part. If you are just presented with two sets of figures like this, the task of comparison is not easy.

Table 10   Heights of men presented in two groups

Group 1 (height/cm)					Group 2 (height/cm)				
146	160	166	171	178	145	160	166	171	179
149	161	166	172	178	147	160	167	172	179
150	162	167	172	179	150	161	167	173	180
152	162	168	173	180	150	162	167	173	181
152	163	169	174	181	151	162	167	174	182
154	163	169	175	181	154	162	169	174	184
156	164	169	175	183	156	164	169	175	186
157	165	170	176	187	156	165	169	176	186
157	165	170	177	188	158	165	170	176	189
159	166	171	178	190	159	165	171	177	190

The task is made much easier if both sets of data are represented in a histogram using the *same* classes.

Figure 18 shows the two histograms on one pair of axes. The unbroken lines show Group 1 data and broken lines show Group 2 data: where only

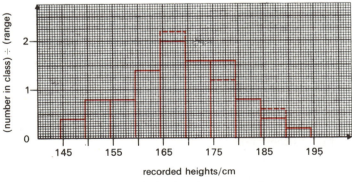

Figure 18   The histograms of the heights of two groups of men.

a full line appears in a class, the two histograms are identical. You can now carry out a quick visual comparison.

What can you deduce from just looking at Figure 18? Are there any overall differences between the two groups?

You can see that the numbers of men are exactly the same in all but three classes and there is no obvious overall difference. It is not possible, for instance, to state that the men in one group tend to be smaller or taller than those in the other group.

When you draw histograms in order to compare different sets of data or samples, you must present the information from each set of data in a similar way.

SAQ 19

SAQ 19

The data in Table 11 represent the heights of a group of fifty women, recorded to the nearest centimetre. Draw a histogram, choosing your classes with a view to comparison with Figure 18.

Table 11

141	152	157	162	168
143	152	157	162	169
144	153	158	163	170
145	154	158	163	171
146	154	159	164	171
146	154	160	164	172
148	155	160	165	173
149	155	160	167	176
150	156	160	168	177
151	157	161	168	177

Compare your histogram with those in Figure 18 and comment on any similarities or differences.

One similarity you may have noticed between Figure 18 and Figure 38 is that the three histograms have very roughly the same shape. If you look at Figures 17 and 35 you will see that they, too, have a similar shape. All the histograms have their highest point around the middle and fall away approximately symmetrically each side of this peak. This shape arises quite commonly when the distribution of a continuous quantity is plotted on a histogram and I shall discuss this shape again in Section 5.

4.5   Prediction using histograms

You have already seen in Section 4.3 that one use of histograms is in prediction. The skirt manufacturer obtained a sample of measurements and then used the sample to predict the sizes of skirts required for the whole potential market. Histograms can also be used for prediction in other manufacturing industries. Figure 19 shows the histogram obtained when forty packets of sugar were selected at random and weighed in the packing room of a refinery. The sugar was to be sold in 1 kilogram (1000 gram) packets. The shape is again of the type I drew your attention to in Section 4.4, but the peak value is not at 1000, nor even very near it,

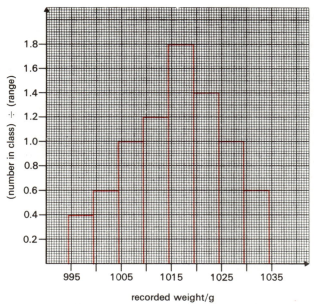

Figure 19   The histogram of the weights of forty packets of sugar.

and yet the packet is called a 1 kg packet.* In fact, only two packets weigh less than 1 kg, while thirty eight weigh 1 kg or more.

Why should a sugar refiner produce '1 kg packets' that contain more than 1 kg of sugar?

An underweight packet may cause the company to be liable for prosecution.

There is always a spread of values in a manufacturing process, however much the manufacturer might wish otherwise! The job of the packers is to ensure that as few as possible of the packets are underweight, while not making the peak of the distribution so high that they are forced to raise the price and perhaps lose sales to their competitors. To do this, packing firms need to know the sort of distribution of weights they obtain as the packets come off the production line. Then they have to decide on a distribution which does not have too many packets below the stated weight, and yet does not have a peak too far above it.

Figure 20 shows three different distributions that might be obtained by adjusting the quantities of sugar put out into 1 kg packets. In Figure 20(a)

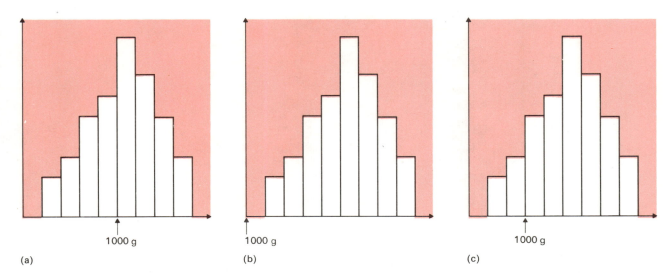

(a)   (b)   (c)

Figure 20   Three histograms of the weights of packets of sugar, obtained by adjusting the packaging process.

* kg is the symbol for kilogram(s): g is the symbol for gram(s).

43

the process has been adjusted so that the peak of the distribution is at the nominal weight of the packet, 1000 grams, but the manufacturer stands in grave danger of being regularly prosecuted! With the distribution shown in Figure 20(b), there is no danger of prosecution, but at quite a cost. Figure 20(c) is an attempt at a compromise, but with five packets out of forty weighing less than 1 kg is the manufacturer very secure even here?

Are manufacturers even secure with a distribution like that in Figure 19? Actually, there is no general rule about the proportion of articles which have to be underweight or undersize before a manufacturer is liable for prosecution, but there are codes of practice specific to different sorts of goods. The local consumer protection department will take a number of things into account, such as the manufacturer's monitoring process and how difficult it is to manufacture the article in question to within defined limits, before recommending prosecution, but clearly manufacturers must always endeavour to keep the proportion of 'under-measurement' articles small.

To be efficient, a manufacturer will first adjust his production process so that the spread in measurement of his processed articles is as small as possible. He will then collect data on his finished articles and modify his control settings so as to provide a suitable compromise between the need to avoid prosecution and the desire to package no more than the stated amount of goods.

One way in which manufacturers can display the data which they have collected is in the form of a histogram. In this case, manufacturers will use predictions from the histogram to ensure that future production does not break the law. They take a sample of their produce, perhaps forty, perhaps fifty, perhaps a hundred items, and from the sample assume that the sizes of all the items will have a very similar distribution to that of the sample. There are well-established quality control procedures based on this kind of sampling to ensure that standards are maintained.

SAQ 20 SAQ 20

Figure 21 shows the distribution of weights of 100 randomly selected nominally 450 gram packets of tea. Suppose prosecution may follow if more than 5% of items are underweight. Comment on the suitability of Figure 21 for making a prediction as to whether prosecution is likely.

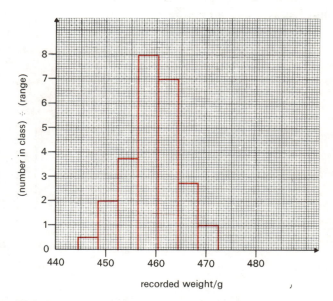

Figure 21   The histogram of the weights of a hundred packets of tea.

4.6 Summary

1 Bar charts are of use when the distribution of discrete quantities is being plotted: histograms are used when the quantity is continuous.

2 The area of a block in a histogram represents the number of items in that class.*

3 The classes chosen in the histogram will depend on the purpose for which the histogram is being plotted. This purpose may be prediction or comparison.

4 The histogram incorporates any modelling suppositions that were made when the data were collected.

The purpose of this footnote is to draw your attention to a use of the word 'histogram' in some texts which may confuse you. There is an alternative way of representing the distribution of a continuous quantity which is called a frequency diagram. In a frequency diagram it is 'number in class' which is plotted on the vertical axis and consequently the classes are always chosen to be all equal in size. Sometimes the word 'histogram' is also used to refer to a frequency diagram and therefore you may meet a diagram called a histogram where the height, not the area, of the block represents the number in the class.

5 THE GAUSSIAN DISTRIBUTION

I have already pointed out that several of the histograms you met in Section 4 were of approximately the shape shown in Figure 22. Histograms that are of this shape that you have already met are:

the distribution of women's waist sizes;
the distribution of men's and women's heights;
the distribution of weights of packets of sugar coming off a production line.

Other histograms which are of a similar shape are:

the distribution of weights of people or of animals;
the distribution of height, girth, etc., of people or of animals;
the distribution of weights or other dimensions of articles coming off a production line.

From this you can see that this sort of shape occurs fairly commonly. Figure 23 shows a curve whose mathematical properties are well known and tabulated; it is called a *Gaussian* or *normal curve*. This curve is symmetrical and is described as bell-shaped.

Looking at Figures 22 and 23, you should notice a distinct similarity in their general *shape*. The histogram of Figure 22 has steps in it and is not the smooth curve of Figure 23, but nevertheless it falls away approximately symmetrically each side of a peak (highest) value, just as the Gaussian curve does.

Because Gaussian curves are well-known mathematical curves whose properties are also well-known, it is often a useful modelling supposition to say that, if you have a sample whose histogram looks rather like Figure 22, it comes from a distribution which is a *Gaussian distribution*— that is, a distribution which is described by a Gaussian curve.

If the weights of a sample of packets of tea have a histogram as shown in Figure 24(a) (which is a repeat of Figure 21) then the distribution of the whole 'population' of packets might be supposed to be Gaussian, as in Figure 24(b). The area of the block shaded red in Figure 24(a) represents the number of packets in the sample between 460.5 and 464.5 g in weight, while the area shaded red in Figure 24(b) represents the *fraction* of packets of tea in the whole 'population' which are between 460.5 and 464.5 g in weight.

If the total number of packets is known then it is easy to use the fraction indicated by the Gaussian curve to calculate the number of packets in any range. For instance, if the fraction of packets between 460.5 and 464.5 g is 0.28 and the total number of packets is 1000, then 1000 × 0.28 = 280 packets have weights between 460.5 and 464.5 g.*

The Gaussian curve of Figure 24(b) can also be used to find the fraction of packets which have weights under 450 g. This is done by finding the area of the region which is shaded grey in Figure 24(b). You will remember from SAQ 20 that this information was not readily available from the histogram of the sample weights. Since the area under any Gaussian curve is tabulated, finding a given area is merely a matter of looking up numbers in a table. Therefore the modelling supposition, that the distribution is Gaussian, has made it possible to deduce information which was not available before.

* *When a number less than one is written in decimal form, it is called a 'decimal fraction' or simply a 'fraction'.*

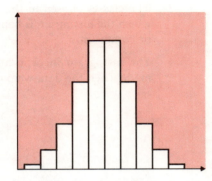

Figure 22   *The shape of several of the histograms in Section 4.*

Gaussian or normal curve

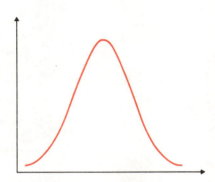

Figure 23   *The shape of a Gaussian distribution curve.*

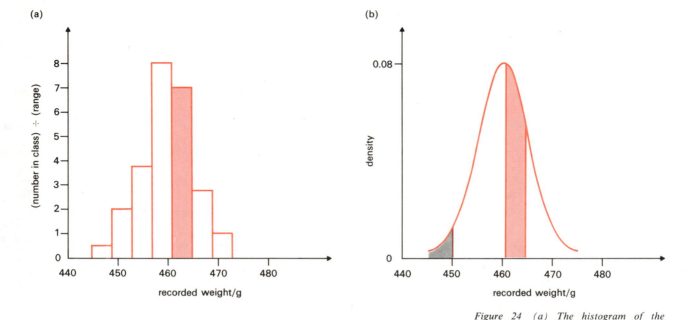

Figure 24 (a) The histogram of the weights of a thousand packets of tea: (b) a Gaussian curve fitted to the distribution of tea packets.

In this course you will not be asked to fit a Gaussian curve to a given histogram, so you do not need to worry about exactly how Figure 24(b) is obtained. The important point to grasp is that once it *has* been obtained it is possible to use it in the manner outlined above.

One important point about Gaussian curves is that they never actually touch the horizontal axis, however far they are extended in each direction. This means that if you use a Gaussian curve to model a distribution which is approximately Gaussian, then your model is going to predict that a very small proportion of items have impossibly small or impossibly large values. For instance, Figure 24(b) suggests that one in a few tens of thousands of packets will have a weight of under 50 g (less than 1/9 the weight marked on the packet)! So while it is useful to model a distribution as Gaussian, it is wise not to accept all mathematical conclusions un-critically, but to remember that it *is* only a model and not reality.

A second important point about Gaussian curves follows on from the fact that the areas of vertical strips under the curve represent the fractions of all the items which lie within those ranges. The point is that the total area must be equal to 1.

 Can you see why this is so?

Look at Figure 25. The area shaded red represents the fraction of packets with weights under 460 g. The unshaded area represents the fraction of

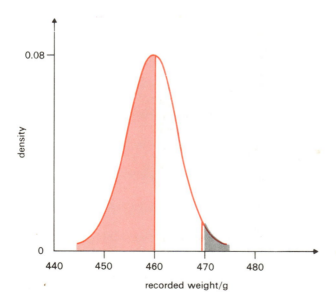

Figure 25 The Gaussian curve of Figure 24(b).

packets with weights between 460 and 470 g and the area shaded grey represents the fraction of packets with weights over 470 g. Since *all* packets lie in one of these three categories the sum of these fractions must be 1, so the total area under the curve must be 1.* (Do not worry about how the area of a shape like this can be calculated—remember that areas have been tabulated so, having been calculated once, they need not be calculated again.)

5.1   Other useful distributions

It is possible to construct histograms that do not approximate to Gaussian distributions. In SAQ 17 you plotted the age distribution of the population of this country and quite clearly that distribution cannot be said to be Gaussian. Other shapes to which histograms may approximate are shown in Figure 26.

Figure 26(a) is a *rectangular distribution* and you should have no difficulty in visualizing how a histogram can approximate to this shape. Figure 26(b) is the shape to which the age distribution of this country approximates. It does not have a special name, but it is again a shape which can be readily

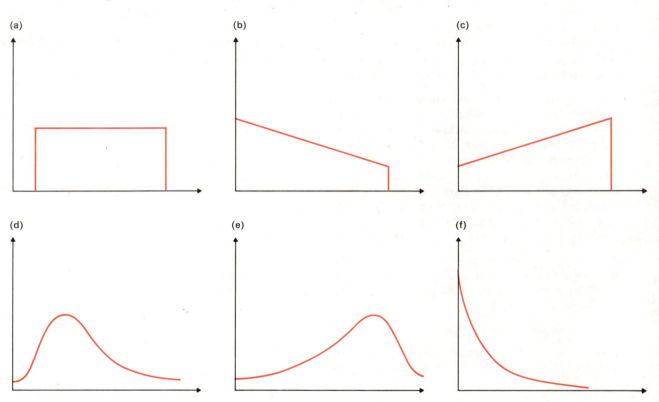

Figure 26   Six shapes to which histograms may approximate.

described in mathematical terms. Figure 26(c) is the 'opposite' of this shape, and you might expect to see it in a country where the birth rate has been declining for some time so that there were more older people than younger ones. Figures 26(d) and (e) represent *skew distributions.* These distributions are not so easy to describe mathematically, since there is not one skew distribution, but many, with different degrees of skewness. This is unfortunate, as sometimes a histogram does approximate to one of these two shapes. Figure 26(f) represents a *negative exponential* distribution.

---

* It is in fact possible to make the total of all the areas of the blocks of a histogram equal to 1. This can be done by dividing the height of each block by the total number of items in the sample. The result is called a normalized histogram.

This distribution often occurs in connection with unpredictable events. For instance, the distribution of the number of phone calls I make or receive at home in a day is of this sort of shape. The negative exponential distribution is also one which can readily be described mathematically.

Just as with a Gaussian curve, it is customary to choose these other curves so that the total area under each of them equals one. The area of any vertical strip under the curve then equals the fraction of items in that range.

5.2   The mean and standard deviation of a Gaussian distribution

One of the main advantages of any Gaussian distribution is that it is characterized by just two numbers called the mean and the standard deviation.

The *mean* (often called the *arithmetic mean*) is the value along the horizontal axis at which the highest point, or peak, of the curve occurs. It is a sort of 'middle value' of the distribution.

mean

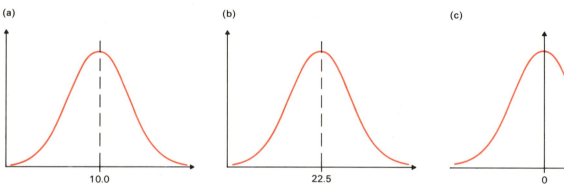

Figure 27   *Three Gaussian distributions with different means.*

In Figure 27(a) the mean is 10.0, in Figure 27(b) it is 22.5 and in Figure 27(c) it is zero. If you are told that you can assume that the distribution of heights of a large group of men is Gaussian and that the mean height of a group of Italian men is 168 cm and the mean height of a group of Swedish men is 174 cm, you can start to draw distributions for these men, as in Figure 28.

Why can you not complete the drawings?

You have not been told anything about how widely the distribution is spread.

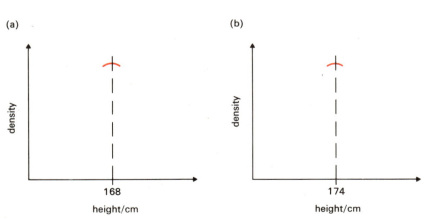

Figure 28   *An attempt to draw the distributions of the heights of (a) Italian and (b) Swedish men, given only the mean of the distributions.*

It is not enough to know a number characterizing the 'middle value' of a Gaussian distribution in order to be able to draw the distribution; a number characterizing the 'spread' is also needed. One such number is called the *standard deviation*. A property of a Gaussian distribution is that the curve falls almost to zero at three standard deviations each side of the mean value. This is shown in Figure 29.

standard deviation

Notice that the curve does not cut off at three standard deviations each side of the mean, but continues falling away towards the horizontal axis. This is because just relatively few items in the distribution may have values outside the main range of values: the range from (mean) – (three deviations) to (mean) + (three standard deviations). The Gaussian distribution allows for 0.27% of the total number of items in the group to be outside that range.

It would be possible to complete Figure 28 provided the standard deviations of the two distributions were known. If the standard deviation for the group of Italian men is 5 cm and that of the Swedish men is 6 cm, Figure 28 can be completed as shown in Figure 30. Notice that for the Italian men the mean is 168 cm and the standard deviation is 5 cm, so nearly all the heights run from $(168) - (3 \times 5)$ cm, which is 153 cm, to $(168) + (3 \times 5)$ cm, which is 183 cm. Similarly, for the Swedish men the heights nearly all run from $(174) - (3 \times 6)$ cm, which is 156 cm, to $(174) + (3 \times 6)$ cm, which is 192 cm.

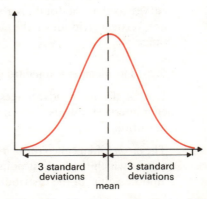

Figure 29 *The Gaussian distribution falls nearly to zero three standard deviations each side of the mean.*

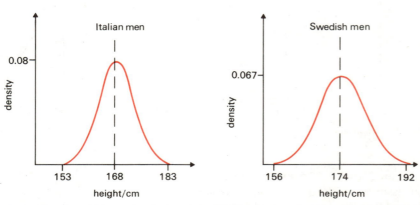

Figure 30 *Figure 28 completed after the standard deviations of the two distributions are known.*

In Figure 30 the peak on the vertical axis of the distribution of the Swedish men is lower than the peak of the distribution of the Italian men. The wider the spread of the distribution, the lower the peak. The reason for this has to do with the area enclosed between the curve and the horizontal axis, which you will remember must always be equal to 1. If the area is to be kept fixed, then clearly as the curve gets more widely spread it must also get lower. In fact the highest value is always approximately 0.4 divided by the standard deviation, and you can check for yourself that this fits with the peak values in Figure 30.

Once the mean and standard deviation of a Gaussian distribution are known, a diagram of the distribution can be sketched. This further reinforces the point that if there are good reasons for believing that any distribution, whose mean and standard deviation are known, approximates to a Gaussian distribution it may be a useful modelling supposition that it *is* Gaussian.

SAQ 21

SAQ 21

Draw diagrams, similar to those in Figure 30, for the following Gaussian distributions:

(a)   A group of men of mean height 170 cm and standard deviation 8 cm.

(b) A selection of packets of flour of mean weight 1.56 kg and standard
deviation 0.02 kg.

SAQ 22

A firm of manufacturers of butter finds that a sample from their packaging
room has a mean weight of 255 g. The nominal weight of the packages is
250 g. They do not want to produce more than 1 % underweight for fear of
prosecution.

Extract any relevant data you may need from Table 12 and say whether
you think they are producing more than 1 % underweight items.

Table 12   Data on the Gaussian distribution of butter packs

when X has this value . . . (weight/g)	. . . area of shaded region in Figure 31 has the value
249	0.001
250	0.006
251	0.023
252	0.067
253	0.159
254	0.309
255	0.500

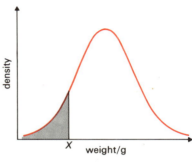

Figure 31   Figure for SAQ 22.

5.3   Summary

A useful modelling step is to say that a distribution which approximates
to a Gaussian distribution *is* in fact Gaussian. This is because a Gaussian
distribution has known properties. For example, the mean of the distribution is the value corresponding to the highest point of the curve and
the curve falls away almost to zero at three standard deviations each side
of the mean. The peak value is approximately equal to 0.4 divided by the
standard deviation. The area under a Gaussian curve is always equal to 1.

6 MEAN AND STANDARD DEVIATIONS

In this section I shall show you how to calculate numerical values for the mean and standard deviation of any set of numbers. If the mean and standard deviation of a set of numbers in a sample is found, these values can be taken as estimates of the mean and standard deviation of the whole 'population' from which the sample was taken.* It is possible to calculate the mean and standard deviation of any set of numbers. The mean will always give a measure of the 'middle value' of the set and the standard deviation will always give a measure of the 'spread' of values in the set. However, only if the set of numbers is a sample from a distribution that is being modelled as a Gaussian distribution is it possible to use the calculated mean as an estimate of the value at which the peak of the distribution occurs and the standard deviation to indicate where the distribution will have fallen away nearly to zero.

6.1 The mean

The mean or 'average' of a set of numbers is found by adding together all the numbers in the set and then dividing by the total number of numbers in the set. Suppose you wish to find the mean of the numbers: 5, 6, 6, 7, 8, 10. Add them up, which gives 42, and then divide by 6, because there are six numbers in the set. The answer is 7, which is the mean value. You are probably already familiar with this process and I shall not discuss it further here, but this method of simply adding all the numbers is rather time-consuming and prone to error, especially when a lot of numbers is involved. There is a 'short cut' method which is to use an *assumed mean*. Appendix 1 shows you how to calculate a mean by this method. This appendix is optional material, but you may like to look through it as the method it illustrates is a useful one.

SAQ 23

Calculate the mean of the following set of numbers:

5	10	13	18
5	10	14	20
6	11	15	21
7	12	16	22
9	13	17	24

SAQ 23

6.2 The square and square root of a number

The *square* of a number is the value obtained when the number is multiplied by itself: the square of 2 is 2×2, which is 4; the square of 5 is 5×5, which is 25. There is a shorthand notation for this, for example: the square of 2 is written as 2^2 (read as 'two squared'); the square of 5 is written as 5^2 (read as 'five squared'), and so on. Thus $2^2 = 2 \times 2 = 4$ and $5^2 = 5 \times 5 = 25$.

square

What are the values of 3^2 and 10^2?

9 and 100, respectively.

---

* Statisticians have found that the standard deviation of a sample is not the best possible estimate of the standard deviation of the distribution. You will meet this point if you study a course on statistics, but for the purposes of this course you may assume that it is a good enough estimate.

The *square root* of a given number is another number which, when multiplied by itself, gives the given number. Thus the square root of 4 is 2, because $2 \times 2 = 4$, and the square root of 16 is 4 because $4 \times 4 = 16$. There is more than one notation for writing square roots; the one I shall use in this unit is one where the square root of 4 is written as $\sqrt{4}$ (read as 'square root of 4', or just 'root 4').*

square root

What are the values of $\sqrt{9}$ and $\sqrt{49}$?

3 and 7, respectively.

6.3 Using a slide rule to find squares and square roots

The slide rule is of use when non-whole numbers are to be squared—$(1.5)^2$ or $(2.75)^2$—or when the number whose square root is to be found is an awkward one—$\sqrt{3}$ or $\sqrt{10}$, neither of which have a square root that is a whole number.

Study Comment

If you have used a slide rule before and know how to use it for finding squares and square roots, you should turn to Section 6.4. Otherwise, now is the time to listen to Side 1 of Disc 1, and to look at your audio disc notes. This recording and associated notes illustrate in simplified form how to read your slide rule and how to use it to calculate squares and square roots. You should also read Sections 2 and 3 in the Slide Rule Book and try the accompanying exercises.

6.4 Calculating the standard deviation

Calculating the standard deviation is a more complicated process than calculating the mean; you must make use of squares and square roots. As an example I shall use the following set of ten values:

7	10
8	11
8	12
9	12
10	13

The mean value of this set of values is 10.

You will remember that the standard deviation is a measure of the 'spread' of a set of numbers. The larger the spread of the numbers, the greater will be the differences between the mean and any of the numbers. If some way is found of 'averaging' these differences then a number which characterizes the spread can be calculated.

This is what is done in finding the standard deviation. The first step is to subtract the mean from each number in the set. Subtracting ten from each number gives:

-3	0
-2	1
-2	2
-1	2
0	3.

---

* *From the rules for multiplying negative numbers, you should remember that two negative numbers, multiplied together will give a positive number. Thus $(-2) \times (-2)$ also equals 4, so the square root of 4 is either $+2$ or -2. Similarly, because $(-4) \times (-4) = 16$, the square root of 16 is either $+4$ or -4. In general, any positive number has two square roots. By the convention the course team will use, $\sqrt{4}$ means the positive square root of 4, and I would have to write $-\sqrt{4}$ if I wanted to denote the negative square root of 4, that is, -2. However, in the context of standard deviation only the positive square root is of any interest, and so I shall concentrate only on the positive square root for the rest of this unit.*

At a first glance it may seem that 'averaging' these differences by finding their mean, as explained in Section 6.1, is the appropriate next step, but of course the mean of these ten differences is just zero. The problem is that the negative values cancel out the positive ones, so if some way could be found of removing the negative values this may solve the problem. The method that is chosen is to square each of the differences. The square of a negative number is a positive number, so this method does indeed remove the negative values. For the set of values I am using the squares are:

9	0
4	1
4	4
1	4
0	9

These squares are then summed, and the sum is 36. The mean of these squares is 36/10 or 3.6.

The standard deviation is the (positive) square root of this value. If you use your slide rule you will find that the standard deviation is 1.9, to two significant figures.

To summarize this procedure for finding the standard deviation:

1   find the mean of the set of numbers;
2   subtract the mean from each number in the set;
3   square each of the numbers resulting from Step 2;
4   find the mean of these squares;
5   find the square root of this mean square.

SAQ 24

SAQ 24

Calculate the standard deviation of the following set of figures, giving your answer to two significant figures:

10	18	23
13	20	24
14	20	26
17	21	27
18	22	27

6.5   Other measures of 'middle value'

You have met the mean as a measure of 'middle value', and you have seen how useful it is to know the mean of Gaussian distribution as this gives the value at which the peak occurs. You have also seen that in the case of the Gaussian distribution the mean and standard deviations are all that are needed in order to characterize it. What about the skew distributions illustrated in Figure 26(d) and Figure 26(e)?

In such cases the mean may not be a useful measure of 'middle value'. Take for instance Figure 32(a), which is a copy of Figure 26(d). The histogram of the distribution of annual incomes of the population of this country would look rather like this since relatively few people have very large incomes. If the mean income was calculated it could give an unfair impression of affluence since the few people with very high incomes would distort it upwards. In such a case a 'middle value' called the median would be more useful than the mean. The *median* is the value such that as many items are above this value than below it. For instance, if the median annual income is £3000, then as many people have an income under £3000 as have an income over £3000. In skewed distributions like Figure 32(a) the median will be lower than the mean.

median

(a)   (b)

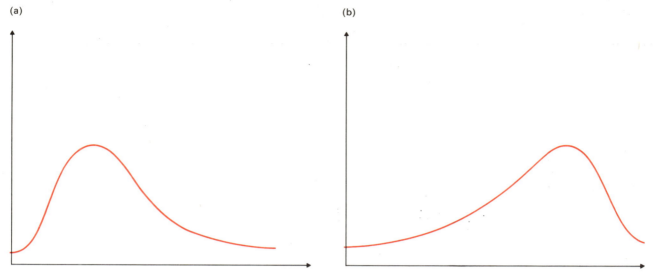

Figure 32   Two skew distributions.

What can you say about the median and the mean of a skewed distribution like Figure 32(b)?

What do you think will be true of the median and the mean of a Gaussian distribution? (Remember that the distribution is symmetrical.)

For a skewed distribution like Figure 32(b), the median will be higher than the mean.

For a Gaussian distribution the 'mean' and the 'median' are equal to each other.

Unfortunately, the common word 'average' can be used to refer either to the mean or the median, and one way in which people can 'bend facts' to suit their case is to select the value that best suits their purpose. There can be a difference between the 'average earnings per man' (which is the mean) and what the 'average man earns' (which may well be the median). Next time you hear or read about an 'average' value, you might like to reflect on whether the mean or the median is meant, and whether there is likely to be any difference between them in the case being discussed.

SAQ 25

You are told that the mean age of the population of a country is 25 and the median age is 23. You wish to choose one of the distributions in Figure 26 to model the population. Which one(s) would you consider using, and why?

SAQ 25

6.6   Summary

1   The mean is a measure of the 'middle value' of a set of numbers and is calculated by adding together all the values in the set and dividing this total by the number of items in the set.

2   The standard deviation is a measure of the 'spread' of a set of numbers and is calculated by subtracting the mean from each value in the set, then squaring these differences, finding the mean of the squares and finding the square root of this mean value. Squares and square roots can be evaluated using a slide rule.

3   Some distributions cannot be approximated by a Gaussian distribution because they are not symmetrical. In such cases, the median may be a better measure of the 'middle value' of the distribution than the mean.

SUMMARY OF THE UNIT

Models are *simplified representations* of reality created for a *purpose* (or purposes). The purposes may be *predicting* future trends, *explaining* phenomena or *designing* things. Any distortions, approximations or simplifications are allowed if they successfully advance the intended purpose. Models which lead to predictions by using *measurements* of number of a chosen population are called *empirical models*. Those that predict or explain on the basis of *understanding* are *causal* or *theoretical models*.

Section 1

The *modelling cycle* expresses the fact that in using mathematics for modelling one begins with *reality*, then forms a *model* that is capable of *mathematical treatment*, then does some mathematical calculation and finally *interprets* the results in terms of reality again.

Section 2

The simplest way to express reality in mathematical terms is by *counting*, but clear categories are needed. This need for categories can cause difficulty and may distort reality, just as other models may. Graphically, the number of items in each category can be represented by means of a *bar chart*. The *number of items* in each category is plotted up the *vertical axis*.

Section 2.1

In expressing a number to, say, *three significant figures*, the *three* digits on the left, after any leading zeros, are used. The *next digit to the right* is used to *check* whether the last of the three digits should be *rounded up* or *down*. It will be *rounded up* if this next digit is 5, 6, 7, 8 or 9 and *rounded down* (that is, remain unchanged) if this next digit is 0, 1, 2, 3 or 4. Other conventions are also in general use, but this is the convention used in *Modelling by Mathematics*. This process of expressing a number to a given number of significant figures introduces a *rounding error*.

Section 2.2

Addition, subtraction, multiplication and division of positive and negative whole numbers and fractions are revised in Section 3.

Section 3

A *bar chart* is used to represent the distribution of items which fall into *discrete categories* and a *histogram* is a way of representing the distribution of a *continuous quantity*.

Section 4.1

In order to collect the data to be represented in a histogram it may be necessary to devise a model by making *modelling suppositions*. One of these may be that a particular group of items is *representative* of the whole, so that the choice of the group on which to make measurements will not introduce *bias* into the survey.

Section 4.2

In preparing to draw a histogram, the data are divided into *classes* which suit the purpose for which the histogram is being drawn. The range of data represented by these classes is plotted along the *horizontal axis* of the graph. The *height* of each block is the *number in the class* divided by *the range*. In this way it is the *area* of each block which represents the *number of items* in that class. This representation is of particular value when the classes are of unequal width.

Section 4.3

Histograms can be used for *comparison* (when the same classes should be chosen for both of the sets of data being compared) or for *prediction*.

Sections 4.4 and 4.5

The histograms of samples taken from 'populations' of heights, weights, etc., often approximate in shape to curves called *Gaussian* or *normal* curves (shown in Figure 23). A useful supposition to make is that such a sample comes from a 'population' whose *distribution is Gaussian*. This then allows the use of the known properties of Gaussian curves. The *area* of any *vertical strip* under this curve represents the *fraction* of items which have measurements in this range. Such areas are tabulated. The *total area* under the curve *equals one*. The measurement corresponding to the *peak* of the curve is called the *mean*. A measure of the 'spread' of the curve is the *standard deviation*. A Gaussian curve falls almost to zero at *three* standard deviations each side of the mean.

Section 5

A histogram does not always approximate to a Gaussian curve. Other shapes which may be found are shown in Figure 26.

Section 5.1

The *mean* of a set of numbers can be found by first adding them up and then dividing by the number of items in the set. A quick method of finding the mean of a large set of numbers is given in Appendix 1 (optional).

Section 6.1

The *square* of a number is the result of *muliplying the number by itself*. The notation used to write the square of 6, say, is 6^2. The *square root* of a given number is the *number which when multiplied by itself, gives the given number*. The notation used to write the square root of 9, say, is $\sqrt{9}$. (The numerical value of $\sqrt{9}$ is 3.) The *slide rule* is of use in finding squares and square roots.

Sections 6.2 and 6.3

The *standard deviation* of a set of numbers can be found by: (1) subtracting the mean from each number; (2) squaring the results of Step 1; (3) totalling the results of Step 2; (4) dividing the result of Step 3 by the number of items in the set; (5) finding the square root of the result of Step 4.

Section 6.4

The *median* of a set of numbers is such that as many numbers in the set are larger than it as are smaller than it. It can be a more useful measure of the 'middle value' than the mean when a distribution of measurements is not symmetrical.

Section 6.5

APPENDIX 1

The use of an assumed mean

Suppose you were asked to average the set of numbers: 51, 52, 54, 55.

You would probably notice that these four numbers can be expressed as

(50 + 1), (50 + 2), (50 + 4), (50 + 5)

and that their average is

$$50 + \frac{1 + 2 + 4 + 5}{4}$$

which is $50 + \frac{12}{4} = 53$.

In this case you have used an assumed mean of 50 and it has helped to make the arithmetic easier by allowing you to deal with the small numbers 1, 2, 4 and 5, rather than the large numbers 51, 52, 54 and 55.

Another example is finding the average of: 98, 99, 100, 101.

In this case the average is

$$100 + \frac{(-2) + (-1) + 0 + 1}{4}$$

which is $100 + (-0.5) = 99.5$.

In this case the assumed mean was 100, and once again its use made the arithmetic easier.

Notice that in both cases the assumed mean did not have to equal the actual mean, it was merely a working basis for it. It was also a 'round' number and that made it easy to subtract it from the other numbers.

The real advantage of taking an assumed mean is seen when the mean of many numbers is to be found. Suppose, for example, I wanted to calculate the mean height of the second group of men whose heights were given in Section 4.4 (Table 13, taken from Table 10).

Table 13

145	160	166	171	179
147	160	167	172	179
150	161	167	173	180
150	162	167	173	181
151	162	167	174	182
154	162	169	174	184
156	164	169	175	186
156	165	169	176	186
158	165	170	176	189
159	165	171	177	190

The first step is to find an assumed mean and I shall take 170 as it is about the middle of this set of numbers and is a fairly easy number to subtract from other numbers—easier, say, than 169.

Having chosen an assumed mean, the next task is to subtract this number from each number in the set. For instance the first number in the first column is 145, and 145 − 170 is −25. Continuing in this way:

−25	−10	−4	+1	+9
−23	−10	−3	+2	+9
−20	−9	−3	+3	+10
−20	−8	−3	+3	+11
−19	−8	−3	+4	+12
−16	−8	−1	+4	+14
−14	−6	−1	+5	+16
−14	−5	−1	+6	+16
−12	−5	0	+6	+19
−11	−5	+1	+7	+20

The next job is to total these differences and then divide by the total number of differences, 50 in this case. Notice that the numbers to be added are much smaller than before—which should reduce errors! They are also both positive and negative.

Many of the numbers in the above set occur both in negative and in positive form—both −9 and +9 occur. Any such pair can be deleted since their sum is zero. The remaining numbers are then added.

−25	−10	−4	+1	+9
−23	−10	−3	+2	+9
−20	−9	−3	+3	+10
−20	−8	−3	+3	+11
−19	−8	−3	+4	+12
−16	−8	−1	+4	+14
−14	−6	−1	+5	+16
−14	−5	−1	+6	+16
−12	−5	0	+6	+19
−11	−5	+1	+7	+20

$$-82 + -44 + -7 + +19 + +25 = -89$$

Since the total of these numbers is −89, their mean is −89/50 which is equal to −178/100 or −1.78.

The assumed mean was 170 and I subtracted 170 from each number in the set. Hence, I must add on 170 to obtain the mean of the original set of numbers (Table 13). The actual mean is 170 + (−1.78), which is 168.22.

You will remember that these were heights recorded to only the nearest centimetre in the first place, so rather than quote the mean with two places of decimals, I shall give it to the nearest centimetre as well, that is, as 168 cm.

To summarize this method of calculating a mean of a set of figures using an assumed mean:

1 Look at the set of numbers and choose an appropriate assumed mean.

2 Subtract the assumed mean from each number in the total set.

3 Look for paired positive and negative numbers and delete them.

4 Total the remaining numbers derived in Step 2.

5 Divide the result of Step 4 by the total number of figures in the set.

6 Add the result of Step 5 to the assumed mean to calculate the actual mean, making sure that you take account of whether there is a positive or negative number to be added to the assumed mean.

SAQ 26

Calculate the mean of the figures in Table 14, preferably by the method of taking an assumed mean. (The figures are those used in SAQ 19, where the heights of a group of fifty women were given.)

Table 14 Heights of a group of fifty women

141	152	157	162	168
143	152	157	162	169
144	153	158	163	170
145	154	158	163	171
146	154	159	164	171
146	154	160	164	172
148	155	160	165	173
149	155	160	167	176
150	156	160	168	177
151	157	161	168	177

ANSWERS TO SELF-ASSESSMENT QUESTIONS

SAQ 1

You might build a model house in order to:

(a) Study its appearance from different elevations.

(b) Try different arrangements of furniture, fixtures or lighting.

(c) Experiment with different colour schemes.

(d) Show its relationship with existing trees (which would also have to be modelled).

(e) Persuade a customer to buy the real house.

(f) Decide where to place solar panels on the roof.

(g) Measure the daylight illumination in the various rooms.

SAQ 2

(a) Since this projection shows the lines of longitude (which meet at a point at each of the poles) as parallel lines, it distorts shapes and the distortion is greater near the poles than near the equator. Canada, being a northerly country, will therefore be distorted and the north of Canada will be more distorted than the south.

(b) Because distances are distorted, as explained in (a) above, it is not possible to use the scale to calculate large distances accurately on Mercator's projection, particularly if either or both of the points are distant from the equator.

SAQ 3

(a) The purpose of the map is to enable passengers to decide the best route to their destination and to show them the line(s) on which they should travel to reach this destination.

(b) (i) The route between adjacent stations has been simplified so it appears as a straight line or a smooth curve on the map.

(ii) The distances between adjacent stations have been distorted so that, in general, stations on the map are approximately equally spaced along the lines, no matter what their true distances apart are.

(iii) As a consequence of (i), the map does not completely preserve direction. For instance, on the map of the underground, Piccadilly Circus appears to be due east of Green Park, while in fact the true direction is north of east.

(c) If the map is used to give an indication of distances between stations (a purpose for which it is not designed) then it could be misleading. For instance, the map suggests that the distance from Oxford Circus to Tottenham Court Road is longer than the distance from Baker Street to St John's Wood, which is not true.

SAQ 4

Figure 33(a) shows a curve drawn through the data points. From this, the estimates for 1981 and 1991 are 55 million and 57 million, respectively. This is only an approximate curve and Figure 33(b) shows another curve which could equally well be drawn through the set of points in Figure 9. This gives the higher estimates of 56.5 million and 59 million for 1981 and 1991.

From the points shown in Figure 9 alone, there is no reason for assuming that either of Figures 33(a) or (b) is better than the other, but since it is currently not as 'fashionable' to have a large family, Figure 33(a) may give better estimates.

One reason for doubting your estimate is that the two World Wars have had a large effect on the population of Great Britain and may have distorted some of the population figures, thus making them unreliable to use for predicting future population.

SAQ 5

(a) *People-in-the-street*—Depending on the time of day, this method is likely to draw upon different, but unrepresentative, groups of people. During the working day, housewives and retired people are most likely to be asked; during the lunch-hour, office workers; and in the evening, those who are not kept at home because of young children.

(b) *House-to-house survey*—it may be that city dwellers have different opinions from small-town dwellers, so that the group in the town may not be representative of the whole country.

(a)

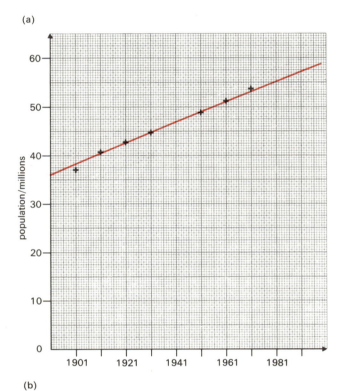

(b)

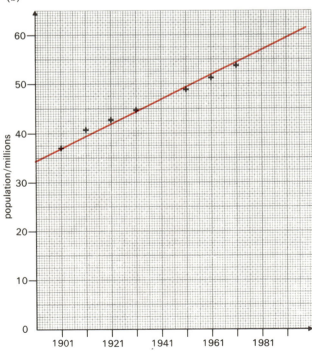

Figure 33 See the answer to SAQ 4.

(c) *A deliberately chosen group*—this is merely selecting *one* characteristic of the potential students and it may not be the most important characteristic; where the students live, or their age, or their family commitments may be much more important, yet this method makes no attempt to distribute these characteristics fairly.

(d) *From the telephone directory*—those who have telephones tend to be better-off and thus the group will not be representative of the whole population.

SAQ 6

The sum is more than the total population of the town because some people may fall into more than one category. For instance, a housewife who has a part-time job may be classified under both 'housewife' and 'earning a living', or a student may register for unemployment benefit in his holidays and thus be classified under both 'unemployed' and 'in full-time education'. It would be necessary to set clear criteria for each category before such anomalies could be removed.

SAQ 7

(a) (i)   1000   (iii) 1250
     (ii) 1300   (iv)   1251

(b) 3.0. The rounding error is 0.05.

(c) 7400. The rounding error is 10.

(d) (i) 0.01.
     (ii) 0.015.

If you express 0.0145 to two significant figures and *then* to one significant figure you will obtain 0.02 for the answer. But 0.0145 is clearly nearer to 0.01 than to 0.02, so it is essential always to express a number *directly* to the given number of significant figures and not to proceed in stages.

(e) 0.001 01. The rounding error is 0.000 003 7.

SAQ 8

Table 15, shows the populations to two significant figures.

Table 15

country	population	approximate population
Sweden	7 495 000	7 500 000
Norway	3 591 000	3 600 000
Denmark	4 585 000	4 600 000
Finland	4 446 000	4 400 000

The bar chart is shown in Figure 34.

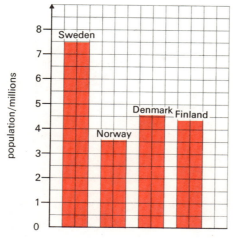

Figure 34   See the answer to SAQ 8

SAQ 9

(a) $467 \times 28 = 13\,076$

(b) $28.5 - 14.2 = 14.3$

(c) $32.3 - 24.5 = 7.8$

(d) $176 \div 8 = 22$

(e) $241 \div 7 = 34.4$ (to three significant figures)

(f) $864 \div 27 = 32$.

SAQ 10

(a) $7 + (-3) = 4$
If you put £7 into your bank account when it was £3 in the red, that is, when the balance was $(-£3)$, the result is a credit balance of £4, hence $7 + (-3) = 4$.

(b) $7 - (-4) = 11$
Clearly, the answer cannot be 3 since $7 - 4 = 3$, and 4 and (-4) are not the same thing. Thinking of bank balances again, if you have £7 and then take away $(-£4)$ you are taking away less-than-zero pounds. That is, you are *putting in* pounds. The result of putting in £4 is to make the new balance £11.

(c) $7 - (-14) = 21$
This is a similar example to question (b).

(d) $-7 - (-8) = 1$
Once again, thinking of bank balances, you have already seen that subtracting a negative amount, $(-£8)$ in this case, is equivalent to putting in (adding) a positive amount, so here you are adding £8 to a balance which is £7 in the red, resulting in a £1 credit balance.

(e) $-8 - 3 = -11$
If you are £8 in the red and withdraw a further £3 you will be £11 in the red.

(f) $-6 - (-5) = -1$
The reasoning here is similar to that given in (d).

(g) $5 \times 3 = 15$

(h) $5 \times (-3) = -15$
The product of a positive number, 5 in this case, and a negative number, (-3) in this case, is always a negative number. (This is a rule for manipulating negative numbers.)

(i) $(-5) \times 5 = -25$
The reason is similar to that given in (h). It does not matter whether the positive or the negative number is the first of the two numbers to be multiplied.

(j) $(-5) \times (-4) = 20$
The product of two negative numbers, (-5) and (-4) in this case, is always a positive number. (This is a further rule for manipulating negative numbers.)

(k) $(-10) \div (-2) = 5$
The quotient of two negative numbers (like the product of two negative numbers) is positive.

(l) $9 \div (-3) = -3$
The quotient of a positive and a negative number (like the product of a positive and a negative number) is negative.

If you feel you need more help with manipulating negative numbers, you should write to the TM281 Course Team, Faculty of Technology, The Open University, Walton Hall, Milton Keynes, MK7 6AA (marking your envelope 'Request for Preparatory Material') for Module O, *Introduction: Fractions and Negative Numbers* of the series *Mathematics Preparatory Material*.

SAQ 11

(a) $\frac{1}{2} + \frac{1}{4} = \frac{3}{4}$

A common denominator is 4, and $\frac{1}{2} = \frac{2}{4}$, therefore

$$\frac{1}{2} + \frac{1}{4} = \frac{2}{4} + \frac{1}{4}$$

$$= \frac{2+1}{4} = \frac{3}{4}$$

(b) $\frac{1}{3} - \frac{1}{4} = \frac{1}{12}$

A common denominator is 12, and $\frac{1}{3} = \frac{4}{12}, \frac{1}{4} = \frac{3}{12}$, therefore

$$\frac{1}{3} - \frac{1}{4} = \frac{4}{12} - \frac{3}{12}$$

$$= \frac{4-3}{12} = \frac{1}{12}$$

(c) $1\frac{3}{4} + 2\frac{1}{8} = 3\frac{7}{8}$

The 'mixed' fractions can first be converted to 'improper' or 'complete' fractions:

$$1\frac{3}{4} = \frac{7}{4} \quad \text{and} \quad 2\frac{1}{8} = \frac{17}{8}$$

A common denominator is 8, and $\frac{7}{4} = \frac{14}{8}$, therefore

$$1\frac{3}{4} + 2\frac{1}{8} = \frac{7}{4} + \frac{17}{8}$$

$$= \frac{14}{8} + \frac{17}{8}$$

$$= \frac{31}{8} = 3\frac{7}{8}$$

(d) $\frac{12}{16} + \frac{5}{20} = 1$

$$\frac{12}{16} = \frac{3}{4} \quad \text{and} \quad \frac{5}{20} = \frac{1}{4}, \quad \text{and} \quad \frac{3}{4} + \frac{1}{4} = 1$$

(e) $\frac{13}{3} - \frac{7}{5} = \frac{44}{15} = 2\frac{14}{15}$

A common denominator is 15, and $\frac{13}{3} = \frac{65}{15}, \frac{7}{5} = \frac{21}{15}$, therefore

$$\frac{13}{3} - \frac{7}{5} = \frac{65}{15} - \frac{21}{15}$$

$$= \frac{65 - 21}{15}$$

$$= \frac{44}{15}$$

$$= 2\frac{14}{15}$$

(f) $5 - 1\frac{4}{5} = 3\frac{1}{5}$

(g) $\frac{1}{2} \times \frac{4}{5} = \frac{2}{5}$

This can be found by multiplying the numerators together (1×4) and then multiplying the denominators together. (2×5). This gives $\frac{4}{10}$, which is $\frac{2}{5}$. Alternatively, 2 can be divided. as shown:

$$\frac{1}{_1 2} \times \frac{4^2}{5} = \frac{1}{1} \times \frac{2}{5} = \frac{2}{5}$$

(h) $1\frac{1}{2} \times \frac{2}{3} = 1$

In multiplication, a mixed fraction should first be converted to an improper fraction. In this case, $1\frac{1}{2} = \frac{3}{2}$, and so

$$1\frac{1}{2} \times \frac{2}{3} = \frac{3}{2} \times \frac{2}{3}$$

$$= \frac{6}{6} = 1$$

or, by dividing,

$$\frac{^1 3}{_1 2} \times \frac{2^1}{3_1} = \frac{1}{1} = 1.$$

(i) $2\frac{1}{3} \times 2\frac{1}{3} = \frac{49}{9} = 5\frac{4}{9}$

Like example (h), this can be found by changing the mixed fractions to improper fractions.

(j) $\frac{4}{3} \div \frac{2}{3} = 2$

In division, the second fraction is inverted and then the resulting fractions are multiplied together, so

$$\frac{4}{3} \div \frac{2}{3} = \frac{4}{3} \times \frac{3}{2}$$

$$= \frac{12}{6} = 2.$$

(k) $3\frac{1}{4} \div \frac{4}{7} = \frac{91}{16} = 5\frac{11}{16}$

The mixed fraction is converted to an improper fraction and the second fraction is inverted, so

$$3\frac{1}{4} \div \frac{4}{7} = \frac{13}{4} \times \frac{7}{4} = \frac{91}{16} = 5\frac{11}{16}$$

(l) $1\frac{3}{4} \div 2\frac{1}{3} = \frac{3}{4}$

The calculation is as follows:

$$1\frac{3}{4} \div 2\frac{1}{3} = \frac{7}{4} \div \frac{7}{3}$$

$$= \frac{7}{4} \times \frac{3}{7}$$

$$= \frac{21}{28} = \frac{3}{4}.$$

If you feel you need more help with manipulating fractions you should write for Module O, *Introduction: Fractions and Negative Numbers* of the series *Mathematics Preparatory Material*.

SAQ 12

There are

$$\frac{5}{9} \times 720 = 400 \text{ girls}$$

$$\frac{1}{8} \times 720 = 90 \text{ children wear spectacles}$$

$$\frac{1}{4} \times 720 = 180 \text{ have blue eyes.}$$

It is not possible to say how many boys do not wear glasses and do not have blue eyes. A first thought might be that $(720 - 400 - 90 - 180)$ boys are in this category, but in fact it is possible that only boys wear spectacles or that none of them do. Equally nature could be quite unfair about blue eyes, giving them all to the girls or all to the boys.

SAQ 13

(a) The sons borrowed one cow from a neighbour, so there were eighteen cows. The eldest son then received nine, the second son received six and the third son received two. This makes a total of seventeen cows, so they were able to return the one borrowed cow!

(b) The eldest son should receive $\frac{1}{2} \times £17\,000$, which is £8500; the second son should receive $\frac{1}{3} \times £17\,000$, which is £5667 (to the nearest pound); and the third son should receive $\frac{1}{9} \times £17\,000$, which is £1889 (to the nearest pound).

61

SAQ 14

(a) A bar chart. Shoe size is a discrete quantity, going up in 'half sizes.

(b) A histogram. The weight of the babies is a continuous quantity which can be measured as precisely as the scales permit.

(c) A bar chart. Each region will have a bar representing its sales and the regions are quite discrete.

(d) A histogram. You may wish to argue that, because there are steps of 1p possible in income, the incomes are discrete, but remember that it is only our banking system which limits money to steps of 1p. Think of petrol prices per gallon—this should help you to see that the concept of money is continuous even if we choose to use it in discrete chunks.

(e) A histogram. The height of the men is a continuous quantity.

SAQ 15

The grouping would be as shown in Table 16.

Table 16

recorded waist measurement/centimetres	number of women
57, 58, 59, 60, 61	21
62, 63, 64, 65, 66	93
67, 68, 69, 70, 71	212
72, 73, 74, 75, 76	141
77, 78, 79, 80, 81	27

The histogram is shown in Figure 35. The decision influences the proportions of size 12 and 16 they manufactured quite strongly, but the other sizes are not much affected, given that the sample would only make an approximate prediction anyway.

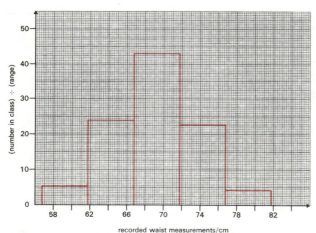

Figure 35 See the answer to SAQ 15.

SAQ 16

The groupings would be as follows (Table 17).

Table 17

recorded waist measurement/centimetres	number of women
53, 54, 55	14
56, 57, 58	38
59, 60, 61	95
62, 63, 64	140
65, 66, 67	123
68, 69, 70	66
71, 72, 73	18
74, 75, 76	6

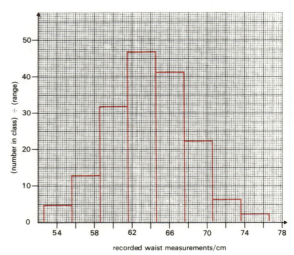

Figure 36 See the answer to SAQ 16.

The histogram is shown in Figure 36. Notice that, as the classes are now 3 cm wide, the number of women in each class must be divided by 3 to give the column height.

SAQ 17

The 'proper' histogram is shown in Figure 37(a). Figure 37(b) shows the results if the 'number in class' were not divided by the range. Notice that the classes were of unequal sizes in the original data, and hence it is imperative to divide each number by the range in order to avoid distortion. One way in which Figure 37(b) could be misleading is that a quick glance might suggest that over twice as many people are aged around 30 than around 20 or that there are more people over 70 than between 60 and 70. Figure 37(a) makes it clear that there are *more* people aged around 20 than around 30. This can be found from the areas of the strips between 19.5 and 20.5 and between 29.5 and 30.5. It also shows that there are more people between 60 and 70 than over 70.

SAQ 18

Box A—women's waist sizes.
          manufactured skirt sizes.

Box B—500 women will be considered, in age-range 16–30, chosen from all the country; data from these women will be taken as representative of *all* possible purchasers; women with measured waist sizes 58–62 cm inclusive will buy size 10, etc.

Box C—manipulation and display of survey data.

Box D—manufacture of skirts of different sizes in the proportions suggested by the survey data.

SAQ 19

The histogram is shown in Figure 38. Notice that because a purpose of the histogram is to compare the data with that already represented in Figure 18, the choice of class is settled—the classes should be the same as those used in Figure 18.

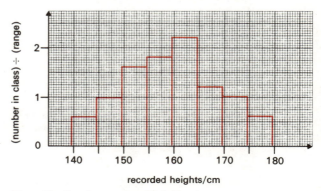

Figure 38 See the answer to SAQ 19.

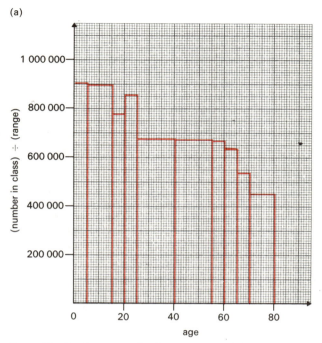

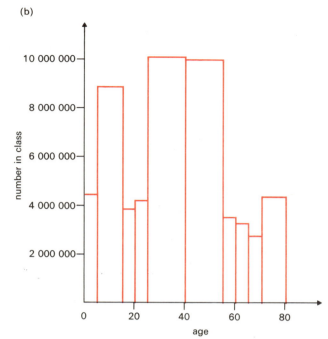

Figure 37   Two histograms for the answer to SAQ 17.

The most obvious difference is that the women in these samples are smaller: the largest class is not 165–169 cm, but 160–164 cm. This of course reflects the fact that women tend, on average, to be smaller than men. However, the general overall shape of Figure 38 is the same as the shape of Figure 18. This is discussed in Section 5.

SAQ 20

In Figure 21 the value of 450 grams is in the middle of a class, rather than being at the lower boundary. Therefore, it is not possible to say how many of the packets in the class 448.5–452.5 grams are in fact under 450 grams. Two packets are known to be underweight because they occur in the class 444.5–448.5 grams and probably rather less than half of the eight packets in the class 448.5–452.5 grams are under 450 grams. so there may be no more than 5 packets (i.e. 5%) underweight, but it is not possible to tell from the figure as it stands. If the person drawing the histogram had been informed of its purpose he would have been able to choose the classes appropriately.

SAQ 21

The diagrams are shown in Figure 39.

SAQ 22

Since 250 g is the nominal weight, the manufacturers are concerned that no more than 1% of packets are under 250 g. Table 12 shows that the area under the Gaussian curve below the value 250 g is 0.006.

If the following modelling suppositions are made:
1   their sample has been selected so it truly predicts the whole;
2   they can characterize the distribution of weights by a Gaussian distribution,

then on this model the proportion of packets under 250 g will be 0.006 (which is 0.6%). Thus on this model they are unlikely to be producing more than 1% of underweight items.

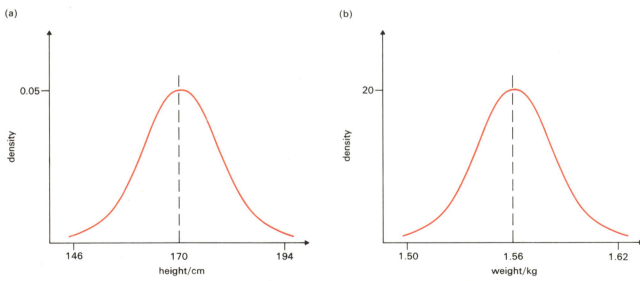

Figure 39   Gaussian distributions for (a) the heights of a group of men and (b) weights of flour packets.

SAQ 23

The mean is 13.4. It is calculated as follows:

5	10	13	18
5	10	14	20
6	11	15	21
7	12	16	22
9	13	17	24

$$32 + 56 + 75 + 105 = 268$$

$$\text{mean} = \frac{\text{sum of all numbers}}{\text{number of numbers}} = \frac{268}{20} = 13.4$$

SAQ 24

The mean of the set of numbers is 20.

Subtracting the mean from each number gives:

-10	-2	3
-7	0	4
-6	0	6
-3	1	7
-2	2	7

and when these are squared, the squares are:

100	4	9
49	0	16
36	0	36
9	1	49
4	4	49

and the sum of the squares is 366. The mean of the squares is therefore

$$\frac{366}{15} = 24.4$$

and the standard deviation is $\sqrt{24.4} = 4.9$, to two significant figures.

SAQ 25

Since the median is lower than the mean, one possibility is a skewed distribution like the one shown in Figure 26(d). The few people in the older age-groups would tend to pull the mean up higher than the median. Another possible distribution is the one shown in Figure 26(b). Once again there are fewer people who are older, but they will tend to pull the mean age up.

SAQ 26

The mean is 159.3, or 159 to the nearest whole number. Since you could have chosen any one of a number of assumed means I cannot show all the possible methods. Below is the calculation if an assumed mean of 160 is chosen:

-19	-8	-3	2	8
-17	-8	-3	2	9
-16	-7	-2	3	10
-15	-6	-2	3	11
-14	-6	-1	4	11
-14	-6	0	4	12
-12	-5	0	5	13
-11	-5	0	7	16
-10	-4	0	8	17
-9	-3	1	8	17

$$-62 + -23 + -3 + 12 + 41 = -35$$

the mean of these differences is $-35/50 = -70/100 = -0.7$. The actual mean is $160 + (-0.7) = 159.3$ cm which is 159 cm to nearest whole centimetre.

Acknowledgements

Grateful acknowledgement is made to the following for illustrations used in this Unit:

Figure 1 National Physical Laboratory. Crown copyright; *Figure 2* Statens Sjohistorik Museum, Wasavarvert; *Figure 3* Hydraulic Research Station; *Figure 5* London Transport; *Figure 9/Table 9* from *Annual Abstracts of Statistics* 1972 reproduced by permission of the Controller of HMSO; *Figure 11* British Railways Board, CEGB and Lever Bros., New York.

2. Linear Models 1

CONTENTS

AIMS

The aims of this unit are:

1   To enable you to construct a model of a simple situation which can be represented by a linear graph and to enable you to draw and use such graphs.

2   To enable you to construct a model of a simple situation where the model can be represented by two linear graphs of which the point of intersection is of interest.

3   To enable you to model a simple situation by determining the appropriate parameters and the relationship between them in order to derive an algebraic equation that describes the model of the situation.

4   To enable you to solve such simple linear algebraic equations.

OBJECTIVES

When you have finished this unit you should be able to:

1   Distinguish between true and false statements concerning, or explain in your own words the meaning of, the following terms:

allowable operations	interpolation
Cartesian axes	linear equation
Cartesian co-ordinates	linear model
dependent variable	origin
dimensions	parameter
extrapolation	subject of an equation
function	substitution
gradient	variable
independent variable	

2   Make the necessary suppositions to arrive at a linear model and represent it by a graph. Plot and read from such a graph and find its gradient (SAQs 1, 2, 3, 4, 5, 6) or use it to make predictions (SAQs 8, 9).

3   Model a situation so that the model can be represented by two lines on a graph and thus determine some point of coincidence in the situation (SAQs 10, 11, 26).

4   Plot a graph, given its equation, and recognize the independent and dependent variable (SAQ 12).

5   Model a situation so that the model can be represented by an algebraic equation, determine the parameters in such an equation and the relationship between them, and substitute into the equation (SAQs 14, 15, 16, 17, 18).

6   Solve a linear algebraic equation (SAQs 19, 20, 21, 22, 23, 24).

STUDY GUIDE

Your work for this study week consists of reading Unit 2, *Linear Models 1* and two sections of the supplementary material, *Modelling Themes* and doing some assignment material.

You are expected to be familiar with the material in Unit 1, especially with the discussion of models and modelling, before you start to read the correspondence text of Unit 2. You will also need to have read the paper on Money in *Modelling Themes* before you start Section 3 of the unit and to have looked at the paper on Town Planning before you start Section 7. Whether you read these papers before you start the correspondence text or when you reach the appropriate section is up to you, but do not spend a long time reading Town Planning—you need to get the feel of the subject matter, but do not need to understand it fully.

Full details of the assessment material associated with Unit 2 are in the *Supplementary Material.*

There is a possibility that you may already be familiar with the material contained in Section 6.3, Solving simple equations. You can test if your understanding of this material is adequate for the purpose of the course by attempting the self-assessment question, SAQ 19: if it is, you can skip Section 6.3 altogether. However, there is no need for you to worry if you cannot answer SAQ 19, as we do *not* expect you to know how to solve simple linear equations before you start this unit. There is more than one way of approaching the solving of linear equations and Appendix 1 gives a different approach to the one in Section 6.3. You may like to turn to the appendix if you find the approach in the main body of the unit difficult.

Because it is so important for you to be able to solve linear equations after you have studied this unit, there is also a program available on the Student Computing Service's computer which deals with the solution of these equations. If you have difficulty in answering the self-assessment questions in Section 6.3 you may find it helpful to use this program. Full details are given in your *Computer Supplement.*

When you have finished the unit you can use the objectives and related self-assessment questions and the summaries to check what you are expected to have accomplished.

1 INTRODUCTION

This unit is concerned with a type of model which is easy to work with and which is frequently used called a *linear model*. This type of model can be represented by a straight-line graph or by a linear algebraic equation. Linear models arise, for example, when a supposition is made that something takes place at a constant rate or constant speed. Because they are so readily manipulable, it is often convenient to use them as a simple description of fairly complex situations.

linear model

The unit starts by looking at a linear model of a railway journey and shows that two different, but equivalent, graphs can be used to represent it. It goes on to discuss how such a model can be constructed and there are exercises on drawing and interpreting the graphs. When two trains are involved, two different straight lines can be drawn on the same graph. Each represents the linear model of one train. The point where these two lines intersect shows when and where one train will overtake the other.

The emphasis of the unit then shifts from graphs to algebraic equations, and the unit shows that both a graph and its corresponding equation will give rise to the same conclusions about the situation being modelled.

You will learn how to set up an algebraic equation of a linear model and how to manipulate such an equation in order to solve it. Since an important part of setting up a model is using an appropriate set of units of measurement, there is a discussion of such units, though the topic is treated again later in the course.

2 MODELS REPRESENTED BY GRAPHS WHICH ARE STRAIGHT LINES

2.1 A model of a train journey

Suppose you were travelling by train from Marseilles to Paris and you suddenly saw a sign: 'Paris 500 km'. Your thoughts might be as follows:

> 500 km and I'm due to arrive in about another four hours. Let me see; that's an average of 125 kilometres per hour.

If you had a timetable handy, you might be able to pursue this thought as follows:

> According to the timetable it's about a quarter of an hour till I reach Lyons. I must be a quarter of 125, about 30 kilometres, away.

You have made a model of the situation. You know perfectly well that the train will stop at Lyons and Dijon and that it cannot therefore travel at a *constant* speed all the way to Paris, but you have decided that for your purposes this modelling supposition, that the train travels at constant speed, is satisfactory. Your purposes are simple, being merely to gain some idea of how far you are from your next stop or when you may expect to pass through a town where the train does not stop (a town which is therefore not shown on the timetable) and so on. With such purposes you are not interested in a model which will give you great accuracy. Instead, you are interested in one which will enable you to do a few simple calculations in your head or, perhaps, on the back of an envelope.

> How would this model of a train travelling at constant speed look if you tried to represent it pictorially by drawing a graph?

It is possible to represent the distance travelled from that 'Paris 500 km' sign after equal intervals of time in a table of values. Table 1 shows how this could be done.

Table 1   Data for the time and distance since passing the sign

number of hours elapsed	number of km travelled
0	0
1	125
2	250
3	375
4	500

In this table, numbers occur in pairs, each pair consisting of a number of hours and a corresponding number of kilometres. The table can be expressed briefly by simply writing down the number pairs: (0, 0), (1, 125), (2, 250), (3, 375), (4, 500), where the *first* number in the pair represents the number of hours that have elapsed since passing the sign and the *second* number in the pair represents the number of kilometres that the train has travelled. The order of the numbers in the pair is important and, therefore, such a pair of numbers is called an *ordered pair*.

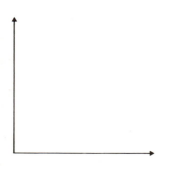

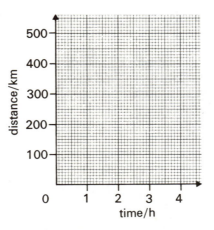

Figure 1   Figure 2

Pairs of numbers can be represented on a graph which has its two axes at right-angles as shown in Figure 1. For any ordered pair, it is conventional to consider the first number as indicating an interval along the horizontal axis to the *right* and the second number as indicating an interval *up* the vertical axis, so that the ordered pair is represented by a single point located a given distance across and a given distance up the graph. You will meet examples of this convention shortly.

This convention was first introduced by the French mathematician Descartes, and hence axes arranged in this way are called *Cartesian axes* and pairs of numbers like those just quoted are called *Cartesian co-ordinates*. If there is no possibility of confusion with other systems of co-ordinates, the word 'Cartesian' is often dropped. I shall do this in the rest of this unit since Cartesian co-ordinates are the only sort I shall use.

Cartesian axes
Cartesian co-ordinates

The axes of a graph should be labelled and then scales chosen for the horizontal and vertical axes so that all the first numbers in the ordered pairs will fit along the horizontal axis and all the second numbers in the ordered pairs will fit up the vertical axis.

In this case, the vertical axis is labelled 'distance/km' and the horizontal axis is labelled 'time/h'.* This is the conventional way of writing 'the number of kilometres travelled since passing the sign' and 'the number of hours elapsed since passing the sign', respectively, on a graph.

Using this convention, how would the vertical axis of a graph be labelled if 'the number of pence an article costs' is being represented along it?

It would be labelled 'cost/p'.

This is necessary because it is *numbers* which are plotted on a graph and the label on the axis tells you what the number refers to. For instance, if the point 500 was plotted on an axis labelled 'distance' you would not know whether it meant 500 metres, 500 kilometres or 500 centimetres, but when the axis is labelled 'distance/km', then 500 means 500 km.

Having drawn and labelled the axes, the next step is to mark them out appropriately. Figure 2 shows this done for the ordered pairs for the train journey to Paris. The first numbers in the ordered pairs run from zero to four and the horizontal axis accommodates them: the second numbers in the ordered pairs run from zero to 500 and they fit along the vertical axis. The number scales along each axis are evenly spaced, but there is no need for each axis to have the same spacing as the other. In this example it would be silly, because the numbers in each pair are of such unequal size.

* h *is the symbol for hour(s).*

9

The point where the axes meet is called the *origin* and has co-ordinates $(0, 0)$: for convenience, it is simply labelled 0.

Figure 3 shows the point whose co-ordinates are $(1, 125)$ plotted on the axes of Figure 2. It corresponds to one interval to the right along the horizontal axis and 125 intervals up the vertical axis. Figure 4 shows the five points corresponding to all the ordered pairs drawn in.

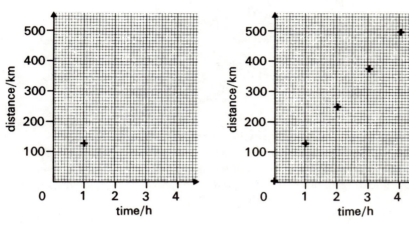

Figure 3   Figure 4

The points are such that a straight line could be drawn through them, but would such a line have any meaning? Figure 5 is a repeat of Figure 4 with a straight line drawn in. What does the point A on the line represent? The broken lines show that this point has co-ordinates $(2.2, 275)$, which means that A represents the point on the journey which is passed 2.2 hours after passing that 'Paris 500 km' sign when the train has travelled 275 km, according to this simple model of its motion. Similar meanings can be given to other points on the line.

What about point B? Point B is represented by the co-ordinates $(5, 625)$ and you will remember that Paris was only 500 km and four hours away. Unless the train has found some way of ploughing across Paris and out the other side, this point must be considered beyond the piece of reality represented by this model. Trying to give meaning to point B would be using the model for a purpose other than that for which it is intended and, as you saw in Unit 1, misleading results are then likely. Of course, if you had seen a sign saying 'Paris 700 km' then a point like point B on a graph of a model of *that* journey would be valid for *that* case.

Figure 6 is a repeat of Figure 5 taking into account that $(4, 500)$ is the extreme point permissible for this model and is a complete graph of the model of the situation.

Earlier, I asked the question: 'How would this model of a train travelling at constant speed look if you represented it on a graph?' You now have the answer; it would look like Figure 6. The important thing to notice is that a model assuming *constant speed* has resulted in a graph which is a *straight line*. It is always the case that straight-line graphs relating distance and time are associated with constant speed. It also turns out that the slope of the straight line is important, because the slope equals the speed. In Figure 6 the graph represents a model of a train travelling at a constant speed of 125 kilometres per hour and its slope is therefore 125.

The word *gradient* is often used instead of slope in mathematics and the gradient of any straight line can be calculated by using the formula

$$\text{gradient} = \frac{\text{a vertical rise of the line}}{\text{corresponding horizontal interval}}$$

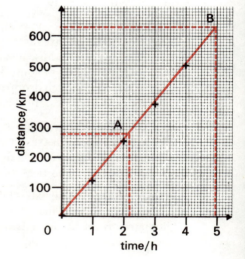

Figure 5

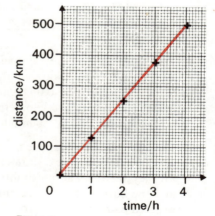

Figure 6

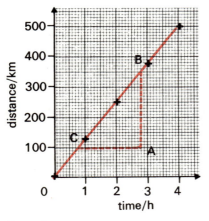

Figure 7

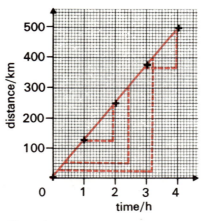

Figure 8

You can see what this means by looking at Figure 7. In Figure 7 the line labelled AB represents 'a vertical rise of the line' and the line labelled CA represents the 'corresponding horizontal interval'. In this case, the vertical rise is 250 and the corresponding horizontal interval is two, and so the gradient is 250/2, which is 125, as expected.

The gradient of a straight line can be found by drawing lines similar to AB and CA (Figure 7) at any pairs of points along the line and finding the vertical rise and corresponding horizontal interval. Before reading any further, check that all of the pairs of lines in Figure 8 give the value 125 for the gradient.

The graph of Figure 6 can be used to answer other questions that may have gone through your mind after passing the sign, such as:

> The timetable shows the train is due to arrive in Dijon in $1\frac{3}{4}$ hours. How far away is Dijon?

This question can be answered by using the graph. From the point $1\frac{3}{4}$ on the horizontal axis draw a vertical line up to meet the line of the graph and then draw a horizontal line across from the line of the graph to the vertical axis. The horizontal line meets the vertical axis at just under 220, so the distance to Dijon should be about 220 km. This is shown in Figure 9. Notice that I have said 'about' 220 km. The modelling supposition that the train travels at constant speed does not justify trying to give results to an accuracy better than to within several kilometres.

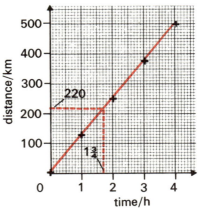

Figure 9

SAQ 1

SAQ 1

Suppose you had been on a different, much slower train when you saw the sign 'Paris 500 km' and that you had then been five hours from Paris according to the timetable. Produce a similar model of this situation and draw the corresponding graph. Check that the gradient of the graph corresponds to the speed of the train. According to your model, how long should it take to get to Dijon (220 km from the sign) on this train?

It may have crossed your mind as you answered SAQ 1 that one of the reasons why this train is slower may not be that it travels any more slowly while it is actually moving, but rather that it stops more often.

11

If this is the case, is the model in SAQ 1 such a simplification that it cannot be considered a useful model?

The answer depends on the purposes you had in mind when preparing the model. A model which is useful for some purposes may not be useful for others. If your purpose was to while away the journey by performing a few simple calculations about where you would be at given times, then the model in SAQ 1 is useful for that purpose. However, to someone preparing a railway timetable and trying to ensure an even and safe flow of trains on the line the model would be very much over simplified. He needs to know where the train stops, and for how long, and also how fast it travels between stations. To him even the model of the motion of the faster train is over simplified. I shall return to this point in Section 2.4.

However, the simplified models of the movement of the two trains are useful not only to the traveller who did the original calculations and plotted the graphs, but they are also of potential interest to other travellers on those trains on the same or other days.

2.2   A different graph of the model

The graph shown in Figure 6 is not the only way of drawing a graph of the situation. To arrive at Figure 6, distances were measured *from the sign*. Another way is to measure distance *from Paris*. Table 2 shows distances from Paris after different intervals of time for the first train.

Table 2   Distance from Paris at hourly intervals

number of hours elapsed	number of km from Paris
0	500
1	375
2	250
3	125
4	0

The ordered pairs, or co-ordinates, of the graph will be: (0, 500), (1, 375), (2, 250), (3, 125), (4, 0); and the graph will be as shown in Figure 10. This is a quite different graph in many ways, but it illustrates the same situation to just the same accuracy.

The major difference between Figures 6 and 10 is that the line in Figure 6 slopes up from left to right while that in Figure 10 slopes down from left to right. In Figure 6 and Table 1, the number of kilometres from the sign is *increasing* as the number of hours *increases*, while in Figure 10 and Table 2, the number of kilometres from Paris is *decreasing* as the number of hours *increases*. This accounts for the different directions of slope of the two lines. Since the model is still one of constant speed, the line in Figure 10 is, as you might expect, straight.

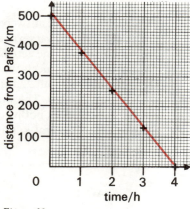

Figure 10

SAQ 2

Use the graph of Figure 10 to answer the following question:

The timetable shows that the train is due to arrive in Dijon 1¾ hours after passing the 500 km to Paris sign. How far is Dijon from Paris?

SAQ 2

Now look at Figure 11 which is a repeat of Figure 10 with the line extended backwards so that it crosses the vertical axis.

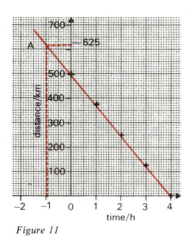

Figure 11

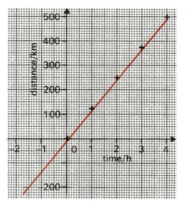

Figure 12

Does this extension have any meaning?

Look at point A, where the distance is 625 and the time is −1. That is, the number of kilometres from Paris is 625 when the number of hours since passing the sign is −1. Put in other words, the number of kilometres from Paris is 625 one hour *before* passing the sign. In this case, the extension of the graph does have meaning, so long as the distance from Paris shown on it is not greater than the total distance the train travelled from Marseilles to Paris (about 800 km). The extension may therefore give useful extra information about the journey. In a similar way, Figure 6 can be extended backwards and will give information about the journey before the train passed the sign, as shown in Figure 12.

In both these cases, negative values of time have been used. You may wonder what 'negative time' means. The point to realize is that I am not talking about negative values of time in the sense of a time interval, because that would lead to nonsensical statements like: 'His journey to work takes −1 hours', but I am using negative values to indicate the number of hours *before* passing the sign. Since positive values of time indicate number of hours *after* passing the sign, this use of negative values is useful and reasonable.

An extension of Figure 6 backwards would not only lead to negative values of time, but also to negative values of distance.

What do you think negative values of distance might mean?

Positive values of distance were used to represent distances on the *Paris* side of the sign. Negative values will represent distances on the *Marseilles* side of the sign, so that a reading of −125 km from the graph represents a distance of 125 km from the sign towards Marseilles, while a reading of +125 km represents a distance 125 km from the sign towards Paris.

To summarize: I have taken one situation, that of a train travelling towards Paris, and I have modelled the situation by supposing it travels at constant speed. I have then represented the model by two graphs. One graph is a plot of the number of kilometres travelled against the number of hours elapsed. The distance increases with time and the graph is shown in Figure 6. The other graph is a plot of the number of kilometres still to be travelled to reach Paris against the number of hours elapsed. The distance decreases with time and the graph is shown in Figure 10. The two graphs are equally valid.

Not only train journeys can be modelled by supposing constant speed of travel; similar methods can be used for car, plane or boat journeys.

SAQ 3

A motorway journey of 475 km takes five hours by car. Make a simple model which could be used by the driver to estimate how far he has travelled. State any modelling supposition(s) you make.

Draw up a table showing how far the car has travelled after successive equal intervals of time have elapsed. Use the table to derive co-ordinates and plot a graph.

From the graph, find:

(a) How far the car has travelled after 1.5, 2.5 and 4.4 hours (give your answers to the nearest 10 km).

(b) After what time it will have travelled 125 and 200 km (give your answers to the nearest 1/10 hour).

SAQ 4

A cross-channel ferry takes two hours to make the 40 km crossing from England to France. Make a simple model of its motion which might be used by a traveller to tell him how far he has travelled. State any modelling supposition(s) you make.

Draw up a table of distances travelled from England after successive equal intervals of time have elapsed and use it to derive co-ordinates and plot a graph.

What is the gradient of the graph?

From the graph, find:

(a) After what time the boat will be 15 km, 25 km and 35 km from England.

(b) How far the boat has travelled after $\frac{1}{4}$ hour and after 1 hour 10 min.

2.3 A model of coal stocks

So far, all the models have related to motion, but similar techniques can be applied to modelling other time-varying situations, such as the stock of a commodity which is being increased at a given rate or which is being drawn at a given rate.

Suppose a householder had 1000 kg of coal delivered to an empty coalhouse and two weeks later estimated he had used 1/10 of it. He wants to have some idea of when he can have a further 1000 kg delivered (his coalhouse holds 1400 kg) and when he is likely to run out if he does not order any more.

Can he use a simple linear model to help him?

Yes. He needs to base it on the modelling supposition that his use of the coal over the first two weeks is typical of his use at any other time, so that if he uses 1/10 of the coal in two weeks he would use it all in $10 \times 2 = 20$ weeks.

This model would lead to the graph shown in Figure 13, which is a straight line. This is a result of supposing that the coal is withdrawn at a constant rate. The line slopes down from left to right because the coal stock is *decreasing* with time.

From the graph it is possible to find out when he could first take delivery of a further 1000 kg. When is this?

After twelve weeks, since he then has only 400 kg left and hence can fit in another 1000 kg.

14

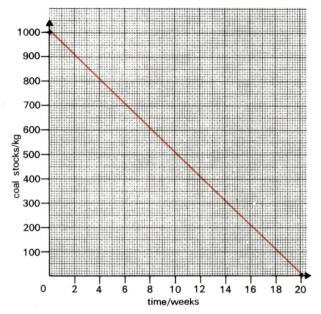

Figure 13

The model also indicates that he will run out of coal after twenty weeks if he has no more delivered, so it suggests he has an eight-week interval during which he can take delivery without fear of running out.

> Suppose that he delays re-ordering and then finds that he runs out of coal after only sixteen weeks. Do you think he will want to use the model represented by Figure 13 again, or do you think he might want to modify it?

Since his purpose is to find out when to re-order and when he is likely to run out, he does indeed need to modify the model. The model indicates that he will still have coal left after sixteen weeks and yet he finds he has none. Clearly, the supposition that the first two weeks are typical is not true. Next time he would be wise to use a model where he uses all the coal in sixteen weeks and work with the results indicated by such a model.

SAQ 5

SAQ 5

Draw the graph he would obtain if he uses this model of 1000 kg lasting sixteen weeks. When can he first take delivery of a further 1000 kg on this model?

2.4 Preparing train timetables

I mentioned in Section 2.1 that the model used for the train journey would not give precise enough results for the purposes of preparing a train timetable, so I am going to return to the situation to see how it can be modelled in a different way.

First, I should point out that it is really not possible to make any model other than the one made previously if all the information available happened to be that Paris is 500 km away and the train will arrive there in four hours. Once you know that the train stops at Dijon and Lyons you are in a position to prepare a model taking such stops into account.

Some of the information is contained in the timetable: for instance, the timetable might tell you that the train stops at Dijon and Lyons for seven minutes each. It would also tell you when you arrive at and leave both of these places as well as when you arrive at Paris. It might also give the distances between Lyons and Dijon and Dijon and Paris.

	Lyons	arr.	16.00
190 km		dep.	16.07
	Dijon	arr.	17.37
290 km		dep.	17.44
	Paris	arr.	19.49

15

Since you also know that when you see the sign you are 500 km from Paris and the time is 15.52, you have all the information you need to prepare a model.

> How can all the information be collected together so that it is in a form where a model can be made and a graph drawn?

The best way is to draw up a table again; Table 3 shows this information. Notice that in Section 2.1 it was enough to say that it would take 'about four hours' to reach Paris, but here it is necessary to be more precise.

Table 3 Time elapsed and distance travelled since passing the sign

time elapsed	number of km travelled
0	0
8 min	20
15 min	20
1 h 45 min	210
1 h 52 min	210
3 h 57 min	500

So far I have done no modelling, since all the values in Table 3 are just data. I can plot the points corresponding to Table 3 on a graph, as shown in Figure 14. Notice that I have chosen to mark off the hours at intervals of thirty small squares along the horizontal axis. I have done this because this will make it easier to plot times to the nearest minute.

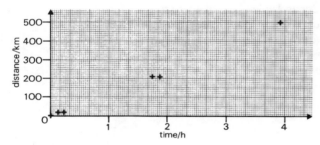

Figure 14

The points do not appear to lie on a straight line. Instead, they seem to go up in steps, the horizontal parts of the steps corresponding to the stops in the stations.

In order to draw a line or lines through the points, it is necessary to make a modelling supposition and, in order to do this, it is necessary to know the purpose of the model. The purposes for which graphs are intended when they are used by people who prepare timetables are to ensure that a slower train will have cleared a section of track before a faster one uses it, that it is safe to slot in an extra goods train or excursion train at a given time of day and so on. For such purposes they have found it adequate to use a model where the trains travel at constant speed between the stations at which they stop.

Remembering that a supposition of constant speed leads to a straight line on a graph of distance against time, Figure 15 shows the completed graph.

Graphs looking like the one in Figure 15 are actually used by people who prepare train timetables, though they are derived in a different way. The train timetabler starts by knowing the journey times for different types of train between pairs of stations and how long the trains should wait at each station. He then draws a graph of the motion and produces the timetable. What we have just done is worked backwards from the timetable to the graph.

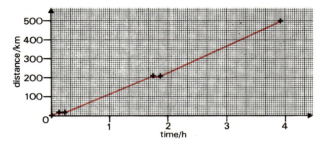

Figure 15

An example of the use of a train timetable graph is as follows. An express train needs to overtake a stopping train somewhere along a stretch of line that has only one track in each direction, except at stations, and it can only do this, therefore, while the stopping train is at a station. The graph of the stopping train's motion is shown in Figure 16(a) and that of the express is shown in Figure 16(b).

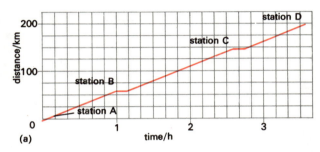
(a)

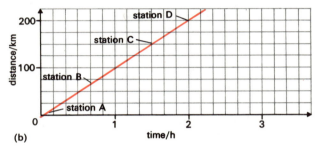
(b)

Figure 16

How can the two trains be timetabled so that the overtaking can proceed smoothly?

Figure 17(a) gives one solution to the problem and Figure 17(b) gives the other. In both cases, the line representing the express train's motion intersects the line representing the stopping train's motion at *horizontal parts*, which represent occasions when the slower train is at a station. The overtaking will therefore be safe. The interval between when the stopping train and when the express leave Station A must be about either twenty-eight minutes or one hour ten minutes.

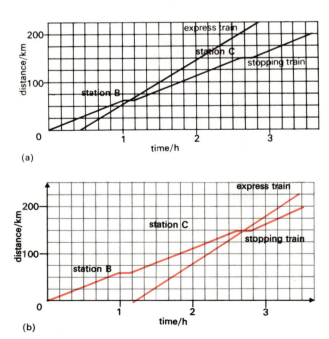

Figure 17

17

Figure 18(a) shows the graph of the motion of a slow train and Figure 18(b) that of a faster train. The two trains are to overtake each other when one is at a station. How can it be done? (Hint: trace Figure 18(b) and move the traced copy of Figure 18(b) over Figure 18(a) until you have found the arrangement you want.)

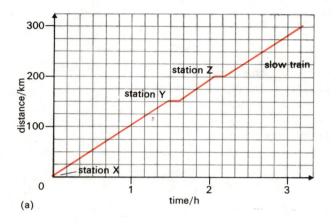

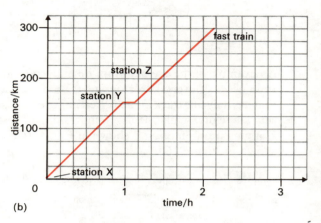

Figure 18

2.5 Summary

In this section you have seen that journeys can be modelled using the modelling supposition of *constant speed* throughout all or parts of the journey. Such a supposition will lead to *straight-line* graphs. The graphs are plotted on *Cartesian axes* with the axes labelled and suitably scaled. The *gradient* or *slope* of the line will give the speed at which the vehicle is travelling.

Straight-line graphs can also apply to other commodities that vary with time at a constant rate.

This self-assessment question is more difficult than those you have met so far. You should be able to answer parts (a) and (b) successfully, but if you do not feel that you are on the road to answering part (c) correctly after about five minutes, do not go on spending time, but turn straight to the answer.

(a) The distance from London to New York is 4800 km, so an aeroplane capable of an average speed of 800 kilometres per hour should be able to do the journey in six hours. Draw a graph of this situation, stating the model of the plane's motion.

(b) Experience shows that the journey averages six hours and forty minutes. A model put forward to explain this is that there is nearly always a headwind as the plane crosses the Atlantic from east to west, which slows the plane down. Calculate the average speed for a journey time of six hours and forty minutes. Draw a graph representing this model *on the same axes* as your previous graph.

(c) What do you think the average wind speed must have been to reduce the average speed of the plane to the value you found in part (b)? What effect would this wind have on the return journey and what do you think the plane's average speed would then be?

3 LINEAR MODELS IN ECONOMICS

Study comment

Before reading this section of the unit you should have read the section in *Modelling Themes* **called Money.**

3.1 The retail price index

In Section 2, I chose to use simple models which could be illustrated by straight-line graphs. I am going to continue that theme in this section, but I shall use examples drawn from statistical data about the retail price index. This will show the way in which a model may have to be altered in the light of new information if a required degree of accuracy is to be achieved.

Table 4 gives the retail price index for the United Kingdom for the six years from 1961 to 1966. These index numbers show how prices changed on a yearly basis. They are calculated in a similar way to the monthly index numbers discussed in Money.

Figure 19 shows these points plotted. Notice that the axes do not start at zero, so that although the price index in 1966 looks about three times as far above the horizontal axis as the index in 1961, the actual value of the index in 1966 (read off the scale on the vertical axis) is only about 20% higher than in 1961. (Just occasionally the decision not to start an axis at zero may be misleading—imagine the effect if Figure 19 showed the sales of 'Brand X' from 1961 to 1966 and it was used in a situation where the numbers along the vertical axis could not be examined very closely.)

What is the advantage of not starting the axes at zero?

To the eye, the points on the graph in Figure 19 appear to lie almost along a straight line, indicating that the retail price index may rise in a linear fashion with time. Given just these six index numbers, there is no obvious reason for assuming any other relationship than the one which appears to emerge from the graph.

Thus, a first attempt at modelling the behaviour of the retail price index with time is to assume that it increases year by year at a constant rate and to represent this model by drawing a straight line like the one in Figure 20. It is

Table 4   Retail price index 1961–66

year	index
1961	94.1
1962	98.0
1963	100.0
1964	103.2
1965	108.1
1966	112.4

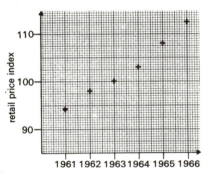

Figure 19

It avoids a large area near the origin which contains no plotted points and enables the points to be plotted on a larger scale on a given sheet of graph paper.

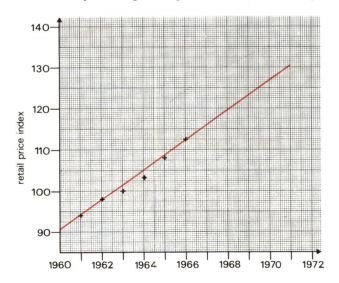

Figure 20

not customary to call the six points plotted in Figure 20 a model, because they simply record the *actual values* of the index. The model is represented by the straight line, the modelling supposition being that the rise of the retail price index is linear with time.

As shown in Figure 21, a curved line can also be drawn through the same points. This represents a different model which, with the limited data given, is not obviously any better or worse than the linear model. At the moment, there is no apparent reason for choosing a more complicated model than the linear one. Economic theory might suggest that a curve represents a more accurate model, but if the purpose of the model is prediction, such a model, like the linear one I have chosen, would have to be tested by comparing its predictions with reality.

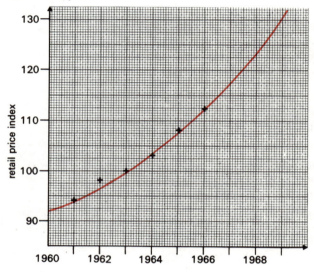

Figure 21

I shall now pause for a moment and ask you to think about the meaning of the line in Figure 20. The retail price index is a number calculated from the prices of various commodities in a given year. Has a line attempting to join these index points any validity? For instance, midway between 1962 and 1963 the line in Figure 20 shows the index to be 99; but in 1962 the index was 98.0 and in 1963 it was 100.0, so was it ever 99?

To help you to answer this question, imagine that the retail price index was calculated monthly instead of annually—or even weekly or daily. Weekly or daily calculations of the index represent extreme cases, but thinking about them, you should be able to see that the retail price index can have valid meaning between the plotted points. Alternatively, you may like to think of a time in 1962 when the goods that cost £100 in 1963 cost £99, and hence a time when the index was 99. The process of reading between plotted points on a graph is called *interpolation*. Thus the graph of Figure 20 is a valid way of modelling the situation so far as interpolation is concerned.

interpolation

No straight line would pass right through *all* the plotted points, of course, and the line in Figure 20 is an attempt at producing a best approximation. One supposition in the model is that the deviation of any point from the line is a small unimportant fluctuation rather than something deeply significant upon which a prediction should depend.

Notice that in Figure 20 I have extended the line in each direction. If the purpose of the model is to predict, then it should reveal something about the likely value of the retail price index after 1966 and it might also be expected to give approximately correct values for the index before 1961.

Reading from the graph in Figure 20 it appears that in 1967 the index should be 115.5 and in 1968 it should be 119. In fact, the values in 1967 and 1968 were 115.2 and 120.6, respectively, so the predictions are not too bad. The differences between the actual values and the values derived from the graph might again be small unimportant fluctuations.

Similarly, the graph would suggest that in 1960 the index should have been 90, while in fact it was 91.

This process of extending a graph beyond the range of the plotted points and then reading from it is called *extrapolation*. So far, the model appears to be giving reasonable predictions when extrapolated, and therefore there are grounds for cautiously considering it to be a useful model for giving future values of the index. These future values are of use to such people as accountants and economists who then have a measure of the probable price level in some years time.

extrapolation

> Use the model, as represented in Figure 20, to write down values for the retail price index from 1969 to 1972, inclusive.

The values you have written down for the period 1969 to 1972 are values which could have been predicted by an economist in 1969 if he was using this model for the behaviour of the retail price index. Often, economic forecasts are made on the basis of models which seem to fit the facts (though not always simple linear models, of course) and important planning decisions are then made on the basis of these models. How would this model have stood up to the test of time?

Figure 22 shows all the actual values of the retail price index from 1955 to 1972 inclusive. The figures in the late sixties and early seventies reflect the fact that inflation started to take a hold at that time and you can see that predictions made in 1969 would have been much too low. In fact, the values appear to be lying on a curve which rises more steeply than the straight line. This indicates that while it is often safe to extrapolate a little it can be dangerous to extrapolate a long way beyond the range of values you have available.

Since prediction of future trends—of price indexes, oil supplies, population, etc.—are continually having to be made, the use of models is an important

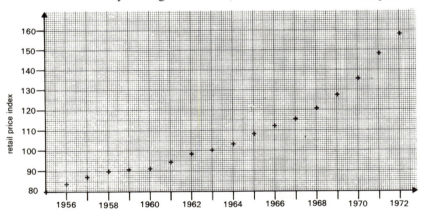

Figure 22

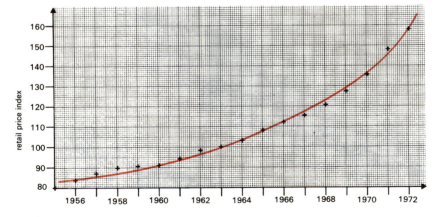

Figure 23

21

activity. As indicated in Unit 1, a good predictive model should not just depend on data; it should be supported by some underlying theory.

When new facts show that the predictions made by a predictive model have fallen short of reality, the next step is to find a new model. Figure 23 shows the points of Figure 22 with a curve, rather than a straight line, drawn through them. Obviously, this curve fits better than any straight line, so it looks as though the curve of Figure 21 would have been a better choice after all. You will meet the subject of modelling a situation with curves later in the course.

SAQ 8

Table 5 shows the population of Great Britain, as given in the ten-yearly census, from 1821 to 1891. Prepare a simple graph of the situation and use it to predict the population in 1901, 1911, 1921, 1961 and 1971.

Would you feel that predictions as far in the future as 1961 and 1971 are justified?

Table 5   Population of Great Britain 1821–91 to the nearest half million

year	population
1821	14 000 000
1831	16 500 000
1841	18 500 000
1851	21 000 000
1861	23 000 000
1871	26 000 000
1881	29 500 000
1891	33 000 000

3.2   Supply and demand

You have already discovered something of what an economist means by 'supply and demand' in your reading of Money in *Modelling Themes* and you may wonder how supply or demand for a particular commodity are determined. One way of estimating demand is by market research surveys, as described in Unit 1. For instance, suppose that a survey of shoppers indicated that the *demand* for raspberries over the whole country would be as shown in Table 6. Figure 24 shows these points plotted on a graph.

Table 6   National demand for raspberries

price/p per kg	number of kg that would be bought/millions
20	30
25	27
30	23
35	20
40	16
45	13
50	10

As you would expect, the demand falls off as the price rises. In this range of prices the points lie approximately on a straight line, and it is possible to model the demand for raspberries by a straight line, as shown in Figure 25. This being the case, you might expect that if the price was fifteen pence per

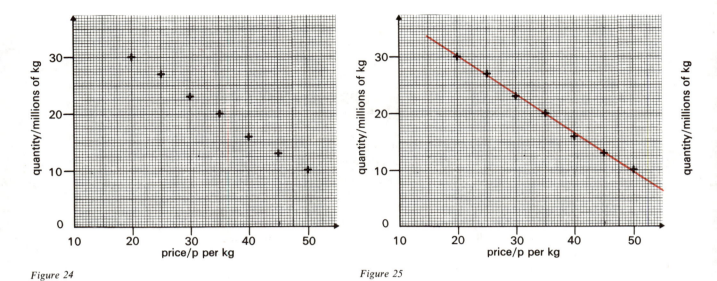

Figure 24

Figure 25

kilogram, then thirty-four million kilograms would be bought, while if the price was fifty-five pence per kilogram, six million kilograms would be bought. This kind of model of the situation indicates an approximate price for a particular level of *supply*.

SAQ 9

SAQ 9

A survey shows that the *supply* of pork carcases as the price varies is as given in Table 7.

Table 7 Variations in the supply of pork carcases with price

price/p per kg	number of kg that would be supplied/millions
80	45
100	49
120	52
140	56
160	60
180	65
200	68
220	72

Prepare a simple graph and use it to predict the number of kilograms which would be supplied at: (a) 70 pence per kg; (b) 130 pence per kg; (c) 240 pence per kg.

4 MODELS WHICH CAN BE REPRESENTED BY A PAIR OF LINES ON A GRAPH

Sometimes it is helpful to devise models which can be represented by a pair of lines on a graph, for the purpose of predicting the point of coincidence of two processes. For instance, models of the supply and demand for a particular commodity can lead to graphs which can be used to predict at what price the demand and supply will be equal. Another example is a case where models about the motion of two vehicles can be used to find the point where one overtakes the other. I shall illustrate this latter case first. The situation is rather like the one you met in the discussion of train timetables in Section 2.4, but the purpose is different here.

4.1 Modelling the motions of two trains

Suppose a commuter train leaves Euston and travels towards Bletchley, and a quarter of an hour later an inter-city train leaves Euston and travels on an adjacent track towards Bletchley. The questions which you might want to answer are how far from Euston the inter-city train passes the commuter train and for how long the commuter train has then been travelling.

The first thing that is needed in order to answer these questions is some knowledge about how fast the two trains are travelling.

Obviously, a commuter train will not travel at a steady speed, but will stop at several stations. A modelling decision will have to be made about how its motion is to be represented and, in order to do this, this purpose must be known. Suppose the purpose is not one which demands very accurate results, so I can take the simplest model and assume that this train travels at a constant speed. Let us suppose the commuter train passes a point eighty kilometres from London after one hour, so that its speed can be taken as 80 km per hour.

For the inter-city train, I can again assume travel at a constant speed, and this is a much smaller simplification, since the inter-city train is not likely to stop more than once, if at all. Using this model, the speed of the inter-city train can be supposed to be 120 km per hour, since it is known to pass a point 120 km from London after one hour.

Using a model of constant speed for both trains, the situation is that a commuter train travelling at 80 km per hour leaves Euston a quarter of an hour before an inter-city train travelling at 120 km per hour. The question

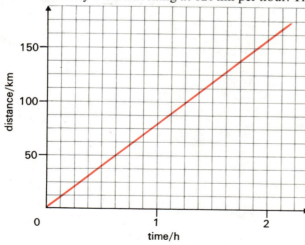

Figure 26

about when the faster train overtakes the slower one can be answered by drawing graphs of each train's motion *on the same axes*.

Figure 26 shows the graph for the commuter train. To plot the graph for the inter-city train on the same axes, it is necessary to remember that at time zero on the graph the inter-city train has not started moving; it does not start to move till a quarter of an hour later. Therefore, the co-ordinates for the inter-city train would be as follows: $(\frac{1}{4}, 0)$, $(1\frac{1}{4}, 120)$, $(2\frac{1}{4}, 240)$.

Are these three points enough to define the line on the graph?

Yes. In fact, just *two* points are enough to fix a straight line. Think about this for a moment until you have convinced yourself that it must be true.

Figure 27 shows the graphs of both train journeys. Notice that the two lines cross.

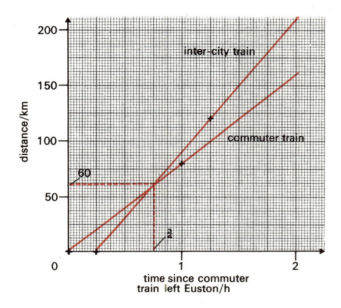

Figure 27

What is the significance of the point where the lines cross?

At this point the distance from Euston is simultaneously the same for both trains—it is the point where the faster train overtakes the slower one. The graphs suggest that this happens instantaneously, but if you have ever been in the situation, you will know that the faster train takes some time to pass the slower one. The graphs suggest that the event happens instantaneously because of a modelling supposition.

Can you see what this supposition is?

The supposition is that the lengths of the trains can be ignored and is implicit in statements like: 'the train is ten kilometres from Euston'. Without this supposition, it would be necessary to make statements like: 'the front buffer of the engine is ten kilometres from Euston', implying that the back buffer is rather less than ten kilometres away. This modelling supposition has led to a result which is not quite true to reality, but quite acceptable in that it will answer satisfactorily the question about where the trains pass each other, bearing in mind that great accuracy is not needed.

The graph indicates that the inter-city train passes the commuter train 60 km from Euston in this model of their motion. The point of intersection is three-quarters of an hour after the commuter train leaves Euston, which means that it is half an hour after the inter-city train leaves Euston.

Whether the results of this simplified model are sufficiently accurate to be useful will depend on the purpose of asking the questions about where and when the faster train overtakes the slower one. If, for example, the purpose is simply to keep a small child amused by telling him or her to begin to watch out for the other train then the results will be adequate.

SAQ 10

SAQ 10

A policeman in a car, the maximum speed of which is 2.5 km per minute, is watching for a stolen car at a motorway interchange. He sees the stolen car go by and joins the motorway one minute later to give chase. The stolen car is moving at its maximum speed of 2.2 km per minute. Use a graph to find out after how long the police car overtakes the stolen car.

Have you had to make modelling suppositions about their motions, and if so, what are they?

4.2 Supply and demand models

The second sort of example I mentioned at the beginning of this section is one of supply and demand. You already know that the supply of a particular commodity is likely to rise as the price rises while the demand for it is likely to fall. If the supply and demand graphs are both plotted on the same axes then it might be possible to derive some information from the point where they cross.

Suppose that the *demand* for a particular variety of apple is as given in Table 8 and the *supply* of the same variety of apple is as given in Table 9. These points can be plotted on a graph, as shown in Figure 28.

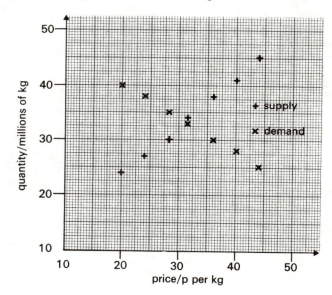

Figure 28

Table 8   Demand for apples

price/p per kg	quantity that would be bought/millions of kg	co-ordinates
20	40	(20, 40)
24	38	(24, 38)
28	35	(28, 35)
32	33	(32, 33)
36	30	(36, 30)
40	28	(40, 28)
44	25	(44, 25)

Table 9 Supply of apples

price/p per kg	quantity that would be supplied/millions of kg	co-ordinates
20	24	(20, 24)
24	27	(24, 27)
28	30	(28, 30)
32	34	(32, 34)
36	38	(36, 38)
40	41	(40, 41)
44	45	(44, 45)

If demand and supply are now both modelled by supposing they change in a linear way with price, then the straight line graphs of Figure 29 result. It is possible to choose the straight lines slightly differently and thus make the intersection of the two lines occur in a slightly different place, so it is necessary to choose lines to go through the scattered points as well as is possible by eye.*

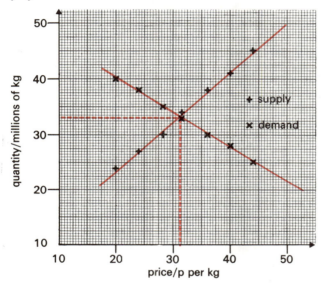

Figure 29

As the lines are drawn in Figure 29, the point of intersection is (31, 33.5). This point corresponds to a price of 31 pence per kilogram and to a quantity of $33\frac{1}{2}$ million kilograms.

What does this point represent?

It is the point where the price is such that the quantity demanded by the consumers just equals the quantity supplied by the producers and it therefore represents the price for this variety of apple at which the suppliers will be able to sell all they have produced (other things remaining equal).

Thus the intersection of the supply and demand graphs for a particular commodity is said to indicate the *equilibrium price* for that commodity.

The term 'equilibrium' is used because, if the model is a fair one, a person selling at below this price would expect to have more customers than he could provide with apples, while selling at above this price would lead him to expect fewer customers than would exhaust the supply of apples actually available.

*There is a mathematical way of determining the 'best' straight line through a set of points, like those from the demand or the supply tables, called 'the method of least squares'. It is beyond the scope of this course to teach you the method, but you should appreciate that mathematical modellers dealing with this sort of data would have techniques other than visual judgement at their disposal, though visual judgement is adequate for these examples.

Table 10 shows the supply and demand for butter. Make any necessary modelling steps and determine the price at which all the butter that is produced will be sold.

Table 10 Supply and demand variations with the price of butter

price/p per kg	quantity demanded/ millions of kg	quantity supplied/ millions of kg
65	30	20
70	28	20.5
75	26	21
80	23	22
85	21	23
90	20	23.5
95	18	25
100	16	26

5 EQUATIONS OF STRAIGHT-LINE GRAPHS

5.1 Relationship between graphs and equations

In this section, I want to introduce you to the idea that a straight-line, or *linear*, graph can be characterized by an equation. Let me take as an example the first graph of the train's journey to Paris after passing the 'Paris 500 km' sign. (Figure 6). This graph was drawn from Table 1, which is repeated as Table 11.

Table 11 Data for the time and distance since passing the sign

number of hours elapsed	number of km travelled
0	0
1	125
2	250
3	375
4	500

Bearing in mind that *any* number multiplied by zero will give the result zero, I hope you can see that if the number of hours elapsed is multiplied by 125 the result is the number of kilometres travelled. This can be written as

number of km travelled since passing sign
= 125 × number of hours elapsed since passing sign

Suppose I let D stand for 'number of kilometres travelled since passing the sign' and T stand for the 'number of hours elapsed since passing the sign'. I can then write the equation describing the data of Table 11 as

$$D = 125 \times T$$

where D and T are just 'shorthand' expressions or symbols, with the meaning ascribed to them above, and stand for numbers.

I want to pause here and look at just what is meant by this equation

$$D = 125 \times T$$

The numbers given in Table 11 are the numbers I used in Section 2.1 to draw the graph in Figure 6. This suggests that there is a very close relation between the equation and the graph. In fact, the relation is so close that it is usual to say that the graph is the graph *of* the equation, or to say that the equation is the equation *of* the graph. The graph 'joins up' the values in Table 11 graphically while the equation 'joins up' the values algebraically. Thus, you have two ways of representing the same model of the train's motion, and they give the same results, to the accuracy to which the graph can be read.

You already know that if you want to find from the graph how far the train will have travelled after half an hour you can draw a line up from $\frac{1}{2}$ on the horizontal axis to meet the line of the graph, and then draw a line across to the vertical axis, as shown in Figure 30. The graph then shows that the distance travelled in half an hour is 62.5 km. The equation will give the same results. If the number of hours is $\frac{1}{2}$, then T is $\frac{1}{2}$, because T represents the number of hours. If $\frac{1}{2}$ is put in the equation instead of T

$$D = 125 \times \tfrac{1}{2}$$

Carrying out the multiplication gives

$$D = 62.5.$$

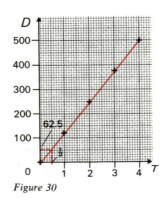

Figure 30

29

Now D stands for the number of kilometres, so the number of kilometres is 62.5, which corresponds to the result obtained from the graph. However, while it is very difficult to read the graph to $\frac{1}{2}$ km there is no problem in using the equation to obtain an answer to $\frac{1}{2}$ km. This suggests that it is possible to obtain more precise results from an equation than from a graph and is one reason for the usefulness of an equation.

I have demonstrated that the graph and the equation give the same result for the distance travelled for just one value of the time, but I could have chosen any other values for the time and demonstrated that the equation and the graph give identical results, to within the accuracy with which you can read the graph. To emphasize this close relationship, I have labelled the axes in Figure 30 D and T, where D and T have the meanings I have assigned to them.

Because the graph of the equation is a straight line, the equation is called a *linear equation.*

From this you will be able to see that the equation

$$D = 125 \times T$$

links values of the distance and the time in such a way that for any value of the time it is possible to calculate the corresponding value of the distance.

For the purposes of my model can I let T have *any* value?

No; values of T greater than four must not be used since the train arrives in Paris and stops after four hours.

This fact, that T must be less than or at most equal to four, in the context of the model, can be indicated by writing $T \leq 4$ alongside the equation. The symbol $\leq$ means 'is less than or equal to', so $T \leq 4$ means that T is less than or equal to four. The equation then looks like this

$$D = 125 \times T \qquad (T \leq 4)$$

This is exactly analogous to stopping the graph at the point (4,500), that is, at Paris.

Since the train started at Marseilles there is also a lower limit for T. You will remember from Section 2.2 that this value will be negative since it indicates a value of time *before* passing the sign. In fact, this lower limit is -2.4, and so T must be 'greater than or equal to' -2.4, in the context of the model. This is written $-2.4 \leq T$, and the two conditions can be combined as shown

$$D = 125 \times T \qquad (-2.4 \leq T \leq 4)$$

5.2   Drawing a graph from a linear equation

If you are given the equation corresponding to a particular model, it is possible to draw the graph of the equation. Suppose you were told that the equation of the model of the motion of the cross-channel ferry you met in SAQ 4 is

$$D = 20 \times T \qquad (0 \leq T \leq 2)$$

where D stands for the number of kilometres from England and T stands for the number of hours that have elapsed since leaving England. The expression $0 \leq T \leq 2$ means that T is greater than or equal to zero and is also less than or equal to two. This just states the fact that the model, and hence the equation, is only valid during the two hours that the ferry is making the crossing.

Study comment

Do not worry that here I am using D and T to stand for something different from their meaning last time. In every new situation, the symbols used are defined to suit the context and some of the symbols will have been used before with other meanings in other situations. Provided you refer to their meanings only within the context of their use there need be no ambiguity.

You could plot the graph of this last equation by writing down values of T and corresponding values of D, as in Table 12, and plotting the co-ordinates.

Table 12   Corresponding values of D and T

value of T	value of D	co-ordinates
0	0	$(0, 0)$
$\frac{1}{2}$	10	$(\frac{1}{2}, 10)$
1	20	$(1, 20)$
$1\frac{1}{2}$	30	$(1\frac{1}{2}, 30)$
2	40	$(2, 40)$

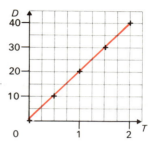

Figure 31

The result is shown in Figure 31, which corresponds to Figure 47 in the answer to SAQ 4, as it should.

5.3   Dependent and independent variables

In both of these cases, D and T are called *variables*. This is because their values can vary (within specified limits) within the models to which they refer. In the equation representing the model of the ferry's motion, T can take any value from 0 to 2 inclusive, and D can take any value from 0 to 40 inclusive. Similarly, in the corresponding graphs, the distances and the times being plotted are called variables. The variable that is plotted along the horizontal axis is called the *independent variable* and the one that is plotted up the vertical axis is called the *dependent variable*. The value the 'dependent variable' takes *depends on* the value of the 'independent variable': for instance, the distance the ferry is from England depends on how long ago it left England.

variable

independent variable
dependent variable

The dependent variable is said to be a *function* of the independent variable. By this is meant that, for any given value of the independent variable, there is just one value of the dependent variable. For instance, in the equation

function

$$D = 20 \times T \qquad (0 \le T \le 2)$$

if you choose any value of T in the range 0 to 2, there is just one value of D which can be calculated from the value of T that you chose.

When D is a function of T then the notation

$$D = f(T)$$

is used (read as 'D equals f of T' or 'D is a function of T'). The letter f and the brackets indicate a function. You will meet functions again later in the course.

Time is almost always the independent variable in any graph that has time as one of the variables, and so time is generally represented along the horizontal axis of a graph (as you have no doubt noticed in the graphs in this unit).

In the context of demand graphs, the consumers were asked 'How much would you buy *at various prices*?' and so price was the independent variable and the quantity depended on the price: the quantity was therefore the dependent variable. Similarly, in graphs of supply, price is again the independent variable.

In equations, the dependent variable is generally the quantity which stands alone on the left hand side of the equation, so that in the equation

$$D = 125 \times T \qquad (-2.4 \le T \le 4)$$

D is the dependent variable.

When a force pulls on a strip of rubber, the rubber stretches. For a particular piece of rubber the equation relating the total length of the rubber to the force pulling it is

$$L = 110 + 5F \qquad (0 \leq F \leq 20)$$

where L is the number of millimetres in the length of the rubber and F is the number of newtons of force applied.* (Note: $5F$ is a shorthand way of writing $5 \times F$.)

(a) What is the independent variable in this case?

(b) What is L when F is: (i) 10; and (ii) 15?

(c) Plot a graph of this equation. Use it to find F when L is: (i) 130; and (ii) 180.

---

*Forces are measured in newtons. To give you an idea of the 'size' of a newton, a force of 10 newtons would be about equivalent to hanging on the rubber one kilogram, the weight of a bag of sugar.

6 MODELS REPRESENTED BY EQUATIONS

In Section 5, you met the equations of some of the graphs presented earlier in the unit: in this section, I want to discuss the use of such algebraic equations in modelling.

6.1 Seeding a lawn

It is possible to estimate the cost of seeding any lawn by preparing a model, writing down the algebraic equation representing the model and then using this general model to calculate the cost for specific lawns and specific seeds. The model and its algebraic equation will therefore be stated in general terms and can be made applicable to many lawns, just by feeding in the appropriate details.

The model will need to incorporate two suppositions. One is that the seed can be spread evenly and accurately so that, if the calculated quantity is bought, no more will be needed to fill patches that are thinly sown or not sown at all because elsewhere the seed was sown too thickly. The other is that it is possible to buy exactly the quantity required (in other words, that the seed is sold 'loose' rather than in packets of fixed size).

To write down the algebraic equation which represents the model it is necessary first to find the quantities on which the cost depends. I shall call these quantities the *parameters*.

parameters

What are these parameters?

The parameters are the *cost per unit amount of grass seed* and the *amount of seed needed for the plot to be seeded*. The total cost of seeding the lawn depends on these two parameters, and the next step is to decide exactly how it depends on them. Should they be added together, or multiplied together, or should one be subtracted from the other or divided into the other? Perhaps the relationship is more complicated; perhaps one should be doubled before it is multiplied by the other, or squared before it is added to the other.

In this particular case, I hope your own experience with this, or a similar domestic problem, will help you to see that

> cost of seeding lawn
> = cost per unit amount of grass seed × amount of seed needed

This 'word equation' states that the two parameters 'cost per unit amount of grass seed' and 'amount of seed needed' are to be multiplied together to give the total cost.

'Amount' is a word which can have different meanings at different times—an amount of meat usually means a weight, an amount of milk a volume, an amount of land an area, and so on. In this case, knowing that grass seed is sold by weight, it is a good idea to replace the word 'amount' in the parameters so that they imply the idea of weight. I shall reword 'cost per unit amount of grass seed' as 'cost per kilogram of grass seed' and 'amount of seed needed' as 'number of kilograms of seed needed'. The word equation then becomes

> cost of seeding lawn
> = cost per kg of grass seed × number of kg needed

I have made the equation more precise by improving on the word 'amount'. The word 'cost' can also be improved on. Does it mean cost in pence or in pounds? It is better to be precise about that as well, and doing so gives

> number of pence to seed lawn
> = number of pence per kg of grass seed × number of kg needed

This word equation is becoming rather cumbersome, but I can define some symbols to stand for the three groups of words.

I shall let

> C stand for the number of pence it costs to seed the lawn;
>
> P stand for the number of pence per kilogram of grass seed;
>
> S stand for the number of kilograms needed.

Then the word equation can be replaced by the symbolic equation

> $C = P \times S$

This equation has exactly the same meaning as the word equation. The symbols P and S are referred to as the parameters in the equation (just like the groups of words they stand for), and C is called the *subject of the equation* because it stands alone on one side of the equation.

subject of equation

In Section 5, I explained the meaning of the word *variable* and showed that symbols may be used to represent variables in an equation. Here I am dealing with parameters and using symbols to represent parameters in an equation. Is there any difference between a variable and a parameter, and if so, what?

There is a difference. A parameter will be a constant in any one application. For instance, P, the number of pence per kilogram of grass seed is a constant. When you have chosen your grass seed you have fixed P. You may have two lawns you want to seed and you may buy two different types of seed and thus have two values for P, but for one application (one lawn) there is one constant value for P. Similarly, there is only one value for S, the number of kilograms needed, for any one lawn. On the other hand, a variable will alter in any one application. For instance, in the train journey to Paris, T, the number of hours elapsed, was not fixed at all; it varied from 0 to 4 as the journey progressed. Similarly, D, the number of kilometres travelled since passing the sign, varied from 0 to 500. Thus, T and D are variables. To decide whether you are dealing with a parameter or a variable ask yourself: 'does it change its value *within* the situation I am modelling?' If the answer is 'yes', it is a variable; if 'no', it is a parameter.

I now need to check that my equation $C = P \times S$ serves the purpose for which I intended it.

> Suppose you were quoted the cost of seed as fifty pence per kilogram and you wanted to seed ten square metres of ground. Would $C = P \times S$ help you or would you need additional information?

Of course, a gardener will measure the area of the plot of land to determine the amount of seed he requires. He wants an equation in terms of 'the number of square metres to be seeded'.

> What additional information will he need?

A parameter which relates, 'the number of square metres to be seeded' to the 'number of kilograms needed'.

This is a parameter that will probably be defined on a seed packet, or can be discovered from gardening books or by asking the seed merchant. Its value may vary slightly with the type of seed or type of soil, but it is certainly a parameter which is closely related to the realities of the situation. In this case, it should take the form of 'the number of square metres covered by one kilogram of seed'. So now I must replace 'the number of kilograms needed', S, by two parameters, 'the number of square metres to be seeded' and 'the number of square metres covered by one kilogram of seed'.

> How should the two parameters, 'the number of square metres to be seeded' and 'the number of square metres covered by one kilogram of seed', be related in order to work out how many kilograms of seed are needed? (Hint: it may help you to think of the problem in terms of simple numbers: if the area of the garden is forty square metres and one kilogram of seed covers twenty square metres, how many kilograms are needed? What have you done with 40 and 20 to obtain your answer?)

The relationship is

number of kilograms needed

$$= \frac{\text{number of square metres to be seeded}}{\text{number of square metres covered by 1 kg of seed}}$$

I have already decided to let S stand for the number of kilograms needed and, if I also let M represent the number of square metres to be seeded and K represent the number of square metres covered by one kilogram of seed, the word equation can be written as the symbolic equation

$$S = \frac{M}{K}$$

> What is the subject of this equation?

The subject stands alone on one side of an equation and does not appear at all on the other side of the equation. The subject of this equation is, therefore, S.

I now have two equations,

$$C = P \times S \quad \text{or} \quad C = PS$$

and

$$S = \frac{M}{K} \quad \text{or} \quad S = M/K$$

Notice that I have given alternative ways of writing these two algebraic equations. The multiplication sign is frequently omitted, as in $C = PS$. Since no other sign is omitted there is no chance of ambiguity. As you saw in Unit 1, Section 3, the division sign is frequently replaced by a horizontal line or by an oblique stroke and you need to be familiar with these notations.

In the two equations

$$C = PS \quad \text{and} \quad S = \frac{M}{K}.$$

you will remember that the problem was that S was not the most realistic parameter to use in calculating C, and hence I found a way of expressing S, using M and K. This new way of expressing S can be put into the equation whose subject is C by a process called *substitution*. Because S equals M/K, this expression M/K can be written instead of S in the equation $C = PS$. The equation then becomes

substitution

$$C = P\frac{M}{K}$$

and M/K has been substituted for S.

SAQ 13

SAQ 13

Express the algebraic equation

$$C = P\frac{M}{K}$$

as a word equation.

Suppose you were told that a certain type of seed costs 110 pence for a kilogram, a plot of ninety square metres is to be seeded and one kilogram of this seed covers fifteen square metres.

How would you calculate the cost of seeding the lawn?

This can be done by substituting into the equation, but this time numbers are substituted in. You are told that

$P = 110$ (the number of pence per kilogram of seed is 110)

$M = 90$ (the number of square metres to be seeded is ninety)

$K = 15$ (the number of square metres covered by one kilogram is fifteen)

and these values can be substituted into the equation to give

$$C = 110 \times \frac{90}{15}$$

which works out as $C = 660$, so the cost is 660p or £6·60 on this model of the cost of seeding a lawn.

You may remember I said at the start of Section 6.1 that the algebraic equation of the model I would arrive at would be a general one which could be made applicable to many lawns by feeding in specific details of the plot of land and type of seed. You have just seen an example of this application of the algebraic equation to one specific case. It can be applied equally well to other cases by the same process of substitution.

SAQ 14

SAQ 14

A certain type of seed costs 120 pence for a kilogram, a plot of 100 square metres is to be seeded and one kilogram of this seed covers twenty square metres. What is the cost of seeding this lawn?

6.2 Units

One very important aspect of modelling is the correct use of units, and I want to discuss this aspect here.

In order to give a value to any physical quantity, such as length or time or weight, it is necessary to quote both a number and a unit. For example, it is not enough to say that a room is three long. The units are missing, so the message does not carry the necessary information. What has been omitted is a unit of length. Units of length are the metre, foot, inch, millimetre, mile, kilometre—some of these are *metric units* (metre, millimetre, kilometre) and some are *British units* (foot, inch, mile). Although Britain has not, at the time of writing, changed completely to metric units, over the life of the course we expect that they will become more commonly used and therefore metric units will be used in the course.

In particular, the course team will use the International System of Units (SI). This is an internationally agreed set of units based on the metric system. In this system:

The base unit of length is the metre.

The base unit of mass is the kilogram.*

The base unit of time is the second.

---

*In everyday speech 'mass' and 'weight' are used interchangeably, the latter being more common. I have used 'weight', for instance, in the seeding a lawn model, and I have done so deliberately: because it is common practice you will be familiar with it. Scientifically, however, mass and weight have somewhat different meanings. You will meet the distinction later in the course when it becomes important.

The kilogram is used rather than the gram because the latter is an inconveniently small base unit.

These three units are symbolized as follows:

unit	symbol
metre	m
kilogram	kg
second	s

In addition, multiples or submultiples of these units can be used. The ones you will probably most commonly meet are:

kilometre (km)	1 kilometre	$= 1000$ metres
millimetre (mm)	1 millimetre	$= \dfrac{1}{1000}$ metre
micrometre (μm)	1 micrometre	$= \dfrac{1}{1\,000\,000}$ metre
gram (g)	1 gram	$= \dfrac{1}{1000}$ kilogram
millisecond (ms)	1 millisecond	$= \dfrac{1}{1000}$ second

From this list, what do you think the prefixes milli- and kilo- denote?

milli- denotes $\frac{1}{1000}$ of a unit

kilo- denotes 1000 times a unit.

In addition, the units 'hour' or 'minute' are sometimes used for lengths of time where they are more appropriate than seconds. Strictly speaking, they are not part of the SI, but in modelling it can be as important to choose units relevant to the problem as it is to choose ones that strictly adhere to the SI. For instance, if you look back to Section 2 where I discussed the train journeys, the hour was the more obvious choice for the unit of time than the SI unit, the second.

The SI unit of speed, metre per second, is *derived* from two of the base units. You have met some other derived units in Section 6.1. For instance, the unit used for area was the square metre, a unit derived from the base unit of length.

The *Course Handbook* gives a list of SI units, with their symbols, for many common quantities like speed and area. You may find this list useful as you study the course.

Quantities that should be measured in units of length, such as distances, are said to have *dimensions* of length. No matter whether you choose the unit of length to be the centimetre, metre, kilometre or even inch or foot, the *dimension* is still length. Similarly, whether speed is measured in kilometre per hour or metre per second, its dimensions are length divided by time. However well you do your modelling, the equations you set up will not be very useful if you make mistakes in your final calculations. One common way in which you may make a mistake is by using different units for quantities with the same dimensions. If you are measuring lengths, say, it will be helpful to keep the various lengths in your calculation in the same units—in metres, or in millimetres, or in kilometres, and so on. As you become familiar with the values of the prefixes listed above, you will get used to writing into your calculations, for example, the 1000 which changes kilometres into metres quite automatically. With times you must also be careful: if you want to do a calculation in seconds, keep to seconds and do not mix in hours or minutes; if you want to use hours, do not use seconds. You should take the same care with derived units, such as speed. If, in a particular example, you use metres for length and seconds for time, you know that speed will be in units of metres per second: all other speeds should then be expressed in metres per second and not kilometres per second or metres per hour.

It is possible to write algebraic equations so that the symbols represent not only the number, but also the units. You will meet equations where the

dimensions

37

symbols represent both the numbers and the units later in the course, but you do not need to know anything about them at the moment other than that they exist.

SAQ 15

SAQ 15

By making a simple model, a home decorator is estimating how many litres of emulsion paint he needs to repaint the walls of a room. His modelling suppositions are:

1   Each coat of paint uses up the same amount of paint.

2   His use of paint is in accordance with the manufacturer's average as stated on the tin.

What parameters should he use to calculate how many litres he needs?

SAQ 16

SAQ 16

(Check your answer to SAQ 15 before answering this question.)

In SAQ 15, what is the relationship between the parameters which will enable him to calculate how many litres he needs?

SAQ 17

SAQ 17

(Check your answer to SAQ 16 before answering this question.)

(a)   In the above example, the decorator now wishes to extend his model to use it to estimate the cost of the redecorating. He makes one more modelling supposition: that he can buy exactly the quantity of paint he needs. What extra parameter or parameters does he need and what is the relationship between all the parameters which will enable him to calculate the cost?

(b)   Rewrite your word equation as an algebraic equation of the situation by choosing suitable symbols to represent the parameters.

(c)   Use your equation to calculate the cost of putting two coats of paint on forty square metres of wall if one litre of the paint covers twenty square metres and costs £1.50.

(d)   Use the symbolic model to calculate the cost of putting three coats of paint on sixty square metres of wall if one litre of the paint costs £1.20 and covers fifteen square metres.

SAQ 18

SAQ 18

A professional decorator is to be asked to paint the walls of a room. His costs include the cost of the paint (as modelled above) and also his labour charges.

(a)   What parameters should be used to model the total cost of painting and how are they related to each other?

(b)   Rewrite the word equation as an algebraic equation of the situation by choosing suitable symbols to represent the parameters.

(c)   Use your equation to calculate the cost of painting a room if he charges £2 an hour, works for twelve hours and spends £10 on materials.

6.3   Solving simple equations

In Section 5, I defined the equation

$$D = 125 \times T$$

as the equation of the graph of Figure 6. If you were asked to find particular values of T given values of D, one way of finding them would be to plot the graph of $D = 125 \times T$ and then read off the values of T corresponding to the given values of D. A much quicker way is simply to use the equation to find the values of T, that is, to solve the equation, and the purpose of Section 6.3 is to teach you how to do this and how to apply the techniques to other similar linear equations.

Study comment

If you can answer all parts of SAQ 19 without any difficulty, turn straight to Section 6.4. If you cannot answer it, *do not worry*: continue reading Section 6.3, the purpose of which is to show you how to solve equations like those in the following self-assessment question.

SAQ 19 SAQ 19

Solve the following equations (that is, find the numerical value of the symbol in each equation):

(a) $2C + 17 = 25$

(b) $15A = 75$

(c) $5P - 12 = 28$

(d) $\dfrac{7U + 2}{5} = 6$

(e) $B - 2 = 2B + 1$

For the purposes of teaching you how to solve linear equations I shall ignore the modelling aspects of the equation $D = 125 \times T$ and concentrate on it simply as an algebraic equation.

If you wished to use this equation, rather than plotting a graph, to find values for D, you need to be given values for T. For instance, if $T = 2$, then

$$D = 125 \times 2$$
$$= 250$$

or if $T = 0.1$, then

$$D = 125 \times 0.1$$
$$= 12.5.$$

It is also possible to find values for T, given values of D. For instance, if D is 375, then

$$375 = 125 \times T$$

and since 3×125 makes 375,

$$T = 3.$$

If D is 300 then

$$300 = 125 \times T.$$

In the earlier case, where D equalled 375, it was easy to spot that T must be three. In this case, it is not so easy to spot a value for T, and so it is necessary to adopt a formal method for the solution of this linear equation.

I shall tell you the rules for solving linear equations and after you have tried a few examples for yourself you should see why the rules are appropriate.

Linear equations are solved by performing one or more of a series of *allowable operations* on them. These operations are:

allowable operations

    Adding the same quantity to both sides of the equation.

    Subtracting the same quantity from both sides of the equation.

    Multiplying both sides of the equation by the same quantity.

    Dividing both sides of the equation by the same quantity.

The quantity referred to may be a number – positive or negative, whole or otherwise (but in the case of division not zero)—or it may be a symbol representing a number or it may be a symbol representing both a number and a dimension.

You may like to think of the process of performing allowable operations on an equation as being rather like the process of changing the weights in the two pans of a balanced pair of scales without disturbing the balance.

Suppose the scales are balanced with six identical weights in each pan. You could subtract the same number (say two) from each pan and the scales would remain balanced. Alternatively, you could divide the number in each pan by, say, three and the scales would stay balanced. If you had more weights you could add the same number of weights to each pan, or multiply the number of weights by, say, two and still the scales would stay balanced.

The scales would go on staying balanced provided you adjusted the weights in both pans in the *same* way. Similarly, an equation goes on being 'balanced', that is, being a valid equation, provided both sides of it are adjusted in the *same* way, as stated in the four allowable operations.

In the case of the equation

$$300 = 125 \times T$$

I can divide both sides of the equation by 125, an allowable operation, and this will cause T to stand alone on one side of the equation.

$$\frac{300}{125} = \frac{125 \times T}{125}$$

Therefore

$$\frac{300}{125} = T$$

The left-hand of this equation can be simplified.

$$\frac{300}{125} = \frac{60}{25}$$

$$= \frac{12}{5}$$

$$= 2.4$$

Therefore

$$T = 2.4$$

Thus, when $D = 300$, $T = 2.4$.

SAQ 20

SAQ 20

Using the equation $D = 125 \times T$, find values for T when:

(a) $D = 250$

(b) $D = 150$

(c) $D = 200$

(d) $D = 400$.

SAQ 21

SAQ 21

Suppose instead the equation was $D = 100 \times T$. Find values for T when:

(a) $D = 50$

(b) $D = 175$.

Suppose instead the equation was $D = 80 \times T$. Find values for T when:

(c) $D = 100$

(d) $D = 240$

Now you should be familiar with the allowable operation 'divide both sides of the equation by the same quantity' and I shall go on to solve other equations by using the other allowable operations. Do not try to attach a meaning to the symbols in the equations: for the moment, as I said earlier, I am concentrating on the techniques for solving linear equations.

Suppose the equation

$$V = U - 20$$

was to be solved, given that $V = 10$. Then

$$10 = U - 20.$$

In order to find a value for U, it is necessary to have U standing alone on one side of the equation. This can be done by adding 20 to both sides of the equation.

$$10 + 20 = U - 20 + 20$$

Therefore

$$30 = U.$$

As another example, suppose the equation

$$S = 2U + 3$$

was to be solved, given that $S = 5$.* Then

$$5 = 2U + 3$$

and again it is necessary to get U standing by itself on one side of the equation. This cannot be done in one step; two steps are needed.

To form $2U + 3$, U was multiplied by 2 and then 3 was added to the result (remember that $2U$ means $2 \times U$). Thus the first step in solving the equation is to take the 3 off again: but it must be taken from both sides in order to keep the equation 'balanced'.

$$5 - 3 = 2U + 3 - 3$$

Therefore

$$2 = 2U$$

This is now similar to the equations you were working with in SAQ 20 and SAQ 21, and both sides can be divided by 2.

$$\frac{2}{2} = \frac{2U}{2}$$

Therefore

$$1 = U \quad \text{or} \quad U = 1.$$

A final example, solve the equation

$$S = \frac{3 + 4V}{2}$$

when $S = 6$. In this case

$$6 = \frac{3 + 4V}{2}.$$

To form the right-hand side of this equation, V was multiplied by 4, then 3 was added to the result and finally the whole thing was divided by 2. Therefore, the first step in getting V to stand alone on one side of the equation is to multiply both sides by 2.

$$6 \times 2 = \frac{2 \times (3 + 4V)}{2}$$

Therefore

$$12 = 3 + 4V.$$

*$2U$ means the same as $2 \times U$. It is an example of omitting the multiplication sign, discussed in Section 6.1.

The next step is to subtract 3 from both sides.

$$12 - 3 = 3 + 4V - 3$$

Therefore

$$9 = 4V$$

Finally both sides can be divided by 4.

$$\frac{9}{4} = \frac{4V}{4}$$

Therefore

$$2.25 = V \quad \text{or} \quad V = 2.25.$$

SAQ 22

Solve the following equations:

(a) the equation

$$V = U - 20$$

    (i) when $V = 5$

    (ii) when $V = 30$.

(b) the equation

$$S = 2U + 3$$

    (i) when $S = 9$

    (ii) when $S = 1$.

(c) the equation

$$S = \frac{3 + 4V}{2}$$

    (i) when $S = 2$

    (ii) when $S = 10$.

(d) the equation

$$L = 110 + 5F$$

    (i) when $L = 160$

    (ii) when $L = 185$.

(e) the equation

$$D = 500 - 125T$$

    (i) when $D = 375$

    (ii) when $D = 0$.

SAQ 23

Use allowable operations to solve the following equations:

(a) $2B + 9 = 7$

(b) $3Q - 11 = 22$

(c) $\dfrac{P + 2}{5} = 6$.

This same method of using allowable operations can be used when the symbol whose value is to be found appears on both sides of the equation. For example, the equation

$$2P + 3 = 5P - 1$$

can be solved using the strategy of getting all the terms with P in them onto one side of the equation and all the terms not including P to the other side. I shall try to get the terms with P in on the right-hand side in this case.

Add 1 to both sides of the equation.

$$2P + 3 + 1 = 5P - 1 + 1$$

which becomes

$$2P + 4 = 5P$$

Subtract $2P$ from both sides of the equation.

$$2P + 4 - 2P = 5P - 2P$$

which becomes

$$4 = 3P$$

This is now in a form you have already met and can be solved by dividing both sides of the equation by 3.

$$\frac{4}{3} = \frac{3P}{3}$$

Therefore

$$1\tfrac{1}{3} = P \quad \text{or} \quad P = 1\tfrac{1}{3}.$$

SAQ 24

Solve the following equations:

(a) $3A + 1 = 2A - 3$

(b) $\dfrac{2Y + 7}{3} = 3Y$

(c) $Y + 1 = \dfrac{2Y + 3}{6}$

Study comment

You should now attempt SAQ 19 at the beginning of this section.

If you have had any difficulty in following this method of solving linear equations, turn to Appendix 1, which gives an alternative presentation and which you may find much easier to follow. When you have studied the appendix, try the self-assessment questions given in the appendix, and then return and check your success by trying some of the self-assessment questions in this section. Alternatively, there is a program available on the Student Computing Service's computer that helps with solving equations like the ones in this unit. You may like to use this program (details are given in your *Computer Supplement*), but do not even think of using it if you have successfully answered SAQ 24 and SAQ 19.

6.4   Summary

In this section, you met some simple situations where the first step in modelling them, after identifying the modelling suppositions, was to find the appropriate parameters and to express them in suitable units. When this had been done the next step was to determine the relationship between the parameters. After symbols had been chosen to represent the parameters, the relationship was written down as an algebraic equation. This equation embodied any suppositions made in modelling the situation and was said to represent the model.

Once an algebraic equation is available it can be used in specific situations by feeding in numerical values for the parameters or variables. In doing this, it may become necessary to solve an equation in which case one of the four allowable operations can be used.

Add the same quantity to both sides of the equation.

Subtract the same quantity from both sides of the equation.

Multiply both sides of the equation by the same quantity.

Divide both sides of the equation by the same (non-zero) quantity.

7   A MODEL USING INEQUALITIES

Study comment

Before reading this section of the unit you should have read quickly through Town Planning in *Modelling Themes*.

A model was made by von Thunen early in the nineteenth century for the purpose of enabling him to determine the types of crop that would be cultivated at different distances from the centre of the town where the crop is sold, and how much producers would be prepared to pay in rent for the land. This model is called von Thunen's agricultural location model. His model has two variables: R, the highest number of pounds the producer would pay in rent per unit area of land; and D, the number of kilometres the land is from the town centre. It also has three parameters: E, the number of kilograms of crop the land yields per unit area; P, the number of pounds profit (selling price less production cost) per kilogram of crop; and T, the number of pounds transport costs per kilometre and per kilogram of crop. These are all related by the equation $R = EP - ETD$

Just look at these symbols for a moment: EP is the number of pounds profit made per unit area of land; TD is the number of pounds transport costs per kilogram of crop and ETD is the number of pounds transport costs for the crop produced by unit area of land. What the equation tells us is that the highest rent per unit area a producer is prepared to pay is his profit on selling the crop less his costs for transporting this crop to where it is sold.

It does not take into account any other factors which may determine this highest rent, such as quality of soil, nor does it allow for the variations in profit which can occur from year to year, so it is a highly simplified model of reality. Nevertheless, it does go some way to fulfilling von Thunen's purpose.

Since the rent I am talking about is the highest rent that a producer will pay I am going to emphasize the point by writing the symbol R_{max} (read as 'R max') instead of R. The subscript 'max' is to remind you that it is the highest or *maximum* number of pounds which is being used as a variable.

Suppose that, for one crop, the values of E, P and T are such that

$$R_{max} = 10\,000 - 1000D \qquad (0 \le D)$$

and for a second crop the values of E, P and T are such that

$$R'_{max} = 12\,000 - 1500D \qquad (0 \le D).$$

I am using the symbol R'_{max} (read as 'R-dashed max') for the second crop to make sure there is no confusion between the rents for the two crops.

Of these two crops, which is more likely to be grown near the town centre?

You should be able to see from the equations that the second crop will be preferred close to the town centre. But how far away is close? This question can be answered by drawing graphs of $R_{max} = 10\,000 - 1000D$ and of $R'_{max} = 12\,000 - 1500D$ on the same axes.

Figure 32 shows the graph of $R_{max} = 10\,000 - 1000D$. You can see that the highest rent the producer is prepared to pay falls off as the distance increases. This is what you would expect; as his transport costs increase, so the highest rent he can pay will fall.

The area below the line in Figure 32 has been shaded. All values of rent within this shaded area are possible rents which might be paid at a given distance from the town centre, but they are all less than the highest value. The region of the graph up to and including the line $R_{max} = 10\,000 - 1000D$ can be represented by the inequality

$$R \leq 10\,000 - 1000D \qquad (0 \leq D, 0 \leq R)$$

where R is the number of pounds the producer can afford to pay in rent per unit area of land (not necessarily the *highest* number of pounds). On the line, $R = R_{max}$, and elsewhere in the shaded region $R < R_{max}$ (the symbol $<$ means 'is less than').

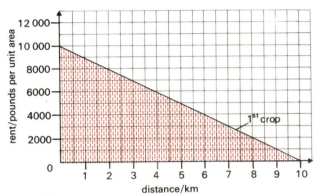

Figure 32

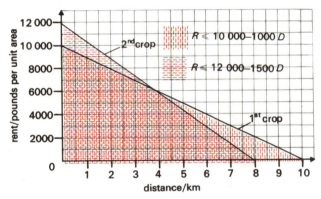

Figure 33

Figure 33 shows Figure 32 repeated, but with $R'_{max} = 12\,000 - 1500D$ drawn in, the new shaded portion being the region

$$R' \leq 12\,000 - 1500D \qquad (0 \leq D, 0 \leq R')$$

You can see that for low values of D (less than $D = 4$), that is, near the town centre, a producer of the second crop can afford to pay higher rents than a producer of the first crop, while for higher values of D (more than $D = 4$), a producer of the first crop can afford to pay higher rents.

Thus von Thunen's first purpose is fulfilled—his model indicates which types of crops will be cultivated at different distances from the town centre. (This method can be extended to more than two crops just by drawing more lines on a graph like the one in Figure 33.)

His second purpose was to find how much producers would be prepared to pay in rent for the land. This purpose is also fulfilled. At $D = 3$, that is, at 3 km from the town centre, the producer of the first crop could pay up to £7000 in rent; the second up to £7500.

Could you have found these two rent values from the equations?

Perhaps the most important point to note about inequalities is that it is often much easier to deal with an equality than it is to deal with an inequality, so it is very common to find that modelling deliberately seeks to introduce equalities. Here I spoke about the *highest* rent being *equal to*, rather than the

Yes, it is simply a matter of solving them given a value for D. For instance, when $D = 3$ then for the first crop

$$R_{max} = 10\,000 - 1000 \times 3$$
$$= 7000.$$

45

rent being *less than or equal to*, a function of the distance from the town centre.

SAQ 25

The relationships between R_{max} and D for two more crops which could be grown are

$$R_{max} = 9000 - 1000D \qquad (0 \le D)$$

and

$$R'_{max} = 15\,000 - 2000D \qquad (0 \le D)$$

By drawing a graph similar to Figure 33, find which of these two crops can pay the higher rent at: (a) 5 km; and (b) 8 km, from the town centre.

At what distance from the centre of the town does the changeover in the two crops occur?

APPENDIX 1

AN ALTERNATIVE METHOD OF SOLVING LINEAR EQUATIONS

The method described in this appendix uses *function boxes*. A function box can be thought of as a device that performs a given operation on the input to it and the result is a unique corresponding output. For instance, the function box may double the input or add ten to it or multiply it by three or subtract five from it, and so on. Figure 34 shows some function boxes with possible inputs and corresponding outputs. Each box is labelled clearly with what it does.

If the input is some number I am calling N, then the output will be given in terms of N. You can see this in the last box in Figure 34, where the input is N, the function is to multiply by two and so the output is $2N$.

The four types of functions I shall use here are addition, subtraction, multiplication and division by a particular number. Figure 34 shows at least one example of each of these types of functions. Also, if I use a letter N to stand for some number then I shall assume that the number N stands for can be a positive or negative whole number or fraction, or it may be zero.

Using function boxes it is possible to see how a 'composite expression' has been formed. Consider for example $2(N + 1)$. The brackets mean that the two multiplies the whole of $N + 1$, so $N + 1$ must be formed first and then this must be multiplied by two. Figure 35 shows the formation of $2(N + 1)$. The number N is fed into the first box where one is added, making the output $N + 1$. This is then fed into the second box where it is multiplied by two, making the overall output $2(N + 1)$.

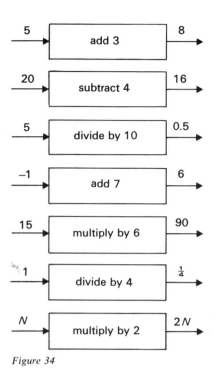

Figure 34

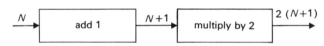

Figure 35

Sketch similar diagrams to Figure 35 to represent

(a) $3N - 1$

(b) $2N + 3$

(c) $\dfrac{N + 4}{7}$

Figure 36 gives the answers.

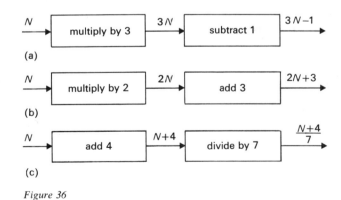

Figure 36

Before I can go on to show you how these boxes can be used to solve equations, I shall also need the 'inverses' of each of the four types of function I am using. By 'inverse' I mean the operation which will 'undo' the operation which the function has done. Let me give a couple of examples.

If 'add four' is the function, then the 'inverse' is the operation which 'undoes' this operation; so the inverse is 'subtract four'.

Similarly, if 'divide by two' is the function, the inverse is the operation which 'undoes' this division; so the inverse is 'multiply by two'.

What are the inverses of:

(a) add one

(b) subtract four

(c) multiply by three

(d) divide by eight?

(a) subtract one

(b) add four

(c) divide by three

(d) multiply by eight.

Now we can go on to see how inverses can be used to solve equations. Suppose the equation to be solved is

$$2(N + 1) = 5$$

You have already seen in Figure 35 how $2(N + 1)$ can be built up. When N is fed in, one is added and then the result is multiplied by two, making the overall output $2(N + 1)$. The equation says $2(N + 1)$ *equals* five, so the output of the final function box must also be five in this case.

To find the value of N that was fed in to make $2(N + 1)$ equal to five, it is necessary to reverse the processes of the function boxes. To do this, the function in each box is replaced by its inverse function; also, the arrows point the other way and the overall *output* of the original function boxes becomes the *input* of the new boxes.

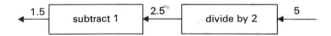

Figure 37

Figure 37 has been drawn with function boxes that are the inverse of those in Figure 35. If five was the output of Figure 35 and five is fed into the first of the new boxes, the output of Figure 37 is 1.5.

So what was N?

N was 1.5.

Let me run through the process again with a new example. Suppose the equation to be solved is

$$3N + 2 = 11.$$

The first thing to be done is to produce a sketch of the function boxes needed to produce $3N + 2$. This is shown in Figure 38. The output is $3N + 2$ which is exactly the same as eleven because the equation says $3N + 2$ equals eleven. To unravel the processes performed on N, the reverse diagram is produced.

Figure 38   Figure 39

In this diagram the inverse functions are used and eleven is fed in. The output of this second combination, as shown in Figure 39, is three. So N must have been three.

Do I need to call the number N? Could I let some other letter represent it— say D, or X, or P, or A or anything I choose?

Yes, *any* letter can represent the number.

Here is one final example: solve

$$\frac{2A - 3}{8} = 5$$

48

The first thing to do is to draw the function boxes. The first function to be performed is to multiply A by two, so this is the first function box.

What is the next function?

Subtract three.

So the second box will subtract three from the output of the first box. Finally, the output of the second box is divided by eight. Figure 40 shows the function boxes.

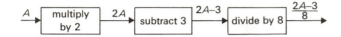

Figure 40

What is the output? Give two values.

It is $\dfrac{(2A - 3)}{8}$ or 5.

The next stage is to produce a reverse diagram and feed five into it. This is shown in Figure 41, where the final output is $21\frac{1}{2}$.

Figure 41

So what was A?

A was $21\frac{1}{2}$.

SAQ 26

SAQ 26

Use the method of function boxes to solve the following equations.

(a) $2P + 1 = 5$ (b) $3X - 7 = 11$ (c) $2(B + 9) = 15$

(d) $\dfrac{3B - 4}{6} = 10$ (e) $3P - 5 = -1$

How do these function boxes work? Let me show you with the following example.

$$\frac{5X + 9}{2} = 12$$

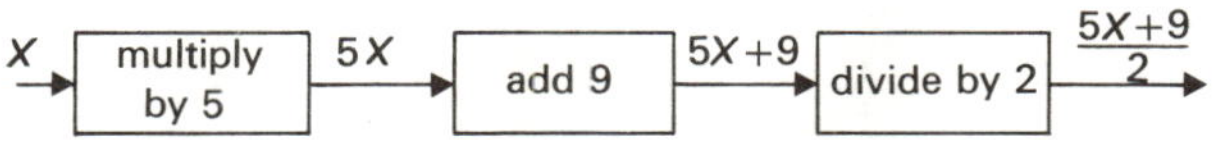

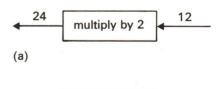

Figure 42

The function boxes are shown in Figure 42. The first step *only* of the reverse diagram is shown in Figure 43(a). In this, the input is being multiplied by two and this input is also the output of Figure 42, so the output of Figure 42 is being multiplied by 2.

$$\frac{5X + 9}{2} = 12$$

Multiply by 2

$$2 \times \frac{5X + 9}{2} = 24$$

(a)

I hope you can see that if you multiply *anything* by 2 and then divide it by 2 you are back where you started. Therefore

$$5X + 9 = 24.$$

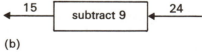

(b)

Figure 43(b) shows the second step. Here 9 is being subtracted from the output of the box in Figure 43(a).

(c)

Figure 43

$$5X + 9 = 24$$

Subtract 9

$$5X + 9 - 9 = 15$$

Once again, adding 9 and then subtracting 9 has no overall effect, so

$$5X = 15$$

Figure 43(c) shows the final step. Here the output of the box in Figure 43(b) is being divided by 5.

$$5X = 15$$

Divide by 5

$$\frac{5X}{5} = 3$$

Multiplying a number by 5 and then dividing it by 5 has no effect on the number, so

$$X = 3.$$

SAQ 27

SAQ 27

Use the method of function boxes to solve the following:

(a) $2X + 3 = 1$ (b) $5X - 16 = 41$ (c) $2(3X - 4) = 52$

(d) $\dfrac{2X + 5}{10} = -1$ (e) $\dfrac{3(X - 2)}{4} = 4$

The next stage is to abbreviate the working by cutting out some of the written stages and 'thinking' them instead. To solve

$$\frac{2B - 3}{6} = 5$$

I would think: 'the last function box would be to divide by 6, so my first function box in reverse is to multiply by 6'.

$$6 \times \frac{(2B - 3)}{6} = 6 \times 5$$

Therefore

$$2B - 3 = 30$$

'Then, the next box in the reverse process would be to add 3'.

$$2B - 3 + 3 = 30 + 3$$

Therefore

$$2B = 33$$

'Finally, the last box would be to divide by 2'.

$$\frac{2B}{2} = \frac{33}{2}$$

$$B = 16\tfrac{1}{2} \text{ or } 16.5$$

From this example, I hope you can see that the function box is a way of making you think of how a function was formed so that you can unravel it in order to solve an equation.

I shall now do one more example where I shall 'think' about the processes of function boxes without actually drawing them. I shall solve the equation

$$9(2X + 1) = 6$$

To obtain $9(2X + 1)$, the functions are:

1 Multiply by two.

2 Add one.

3 Multiply by nine.

To solve the equation, the inverse functions, *in the correct order*, are:

1 Divide by nine.

2 Subtract one.

3 Divide by two.

Starting with the original equation

$$9(2X + 1) = 6$$

carry out the inverse function in the correct order.

Divide by 9

$$9 \times \frac{(2X + 1)}{9} = \frac{6}{9}$$

or

$$2X + 1 = \frac{2}{3}$$

(Note that the fraction has been simplified.)

Subtract 1

$$2X + 1 - 1 = \frac{2}{3} - 1$$

or

$$2X = -\frac{1}{3}$$

Divide by 2

$$\frac{2X}{2} = -\frac{1}{3} \div 2$$

or

$$X = -\frac{1}{6}$$

The solution to the equation is $X = -1/6$.

SAQ 28

SAQ 28

Try to use this method to solve the following equations. Do not worry if you cannot do this; go on using the method of SAQ 27 until, with practice, you find yourself automatically thinking the steps out.

(a) $5X + 11 = -1$ (b) $2X - 9 = -6$

(c) $2(3X + 5) = 19$ (d) $\dfrac{6X - 1}{7} = -1$

SUMMARY OF THE UNIT

A linear model is one which can be represented by a *linear* (straight-line) *graph* or by a *linear algebraic equation*.

Section 1

A linear model of *motion* is made by supposing that a vehicle travels at a *constant speed*. Using *Cartesian axes* and *Cartesian co-ordinates* the graph of a distance travelled against time representing such a model is a *straight line* whose *gradient equals the speed*. The graph can be used to determine how far the vehicle has travelled after a given length of time has elapsed, or vice versa.

Sections 2.1 and 2.2

Linear models also arise when the supposition is made that something happens at a *constant rate*. An example is the linear model obtained by supposing that the same amount of coal is drawn from a stock each week.

Section 2.3

In preparing train timetables, account needs to be taken of the stops the trains make, but it is sufficient to use a model of constant speed between stops.

Section 2.4

In economics, a linear model can be used as a first attempt to predict the future behaviour of, say, the retail price index. Predictions can be made by *extrapolating* the straight line beyond the range of plotted points. *Interpolation* can be used to find values between the plotted points. If an accurate prediction is required it may become necessary to modify the model if it is found that predictions made from it are not sufficiently close to reality. Linear models are also used to estimate *supply* and *demand* of a commodity.

Section 3

The time and place where one vehicle overtakes another can be found by supposing that both vehicles travel at a constant speed and plotting graphs representing this model *on the same axes*. The point where *the lines cross* represents the point where *the overtaking occurs*.

Section 4.1

If linear models of supply and demand are represented on the same axes then the point of intersection represents the price at which *all that is produced will be sold* (this is sometimes referred to as the *equilibrium price*).

Section 4.2

A *linear graph* always has a *linear algebraic equation* associated with it. This equation will have two *variables*—the *independent variable* which is plotted along the *horizontal axis* and the *dependent variable*, plotted on the *vertical axis* and usually the *subject of the equation*. The dependent variable is said to be a *function* of the independent variable, and in linear equations for any value of the dependent variable there is *only one value* which the dependent variable can have. Since the equation represents a model, there may well be *limits on the values* which the variables may take to correspond to the situation being modelled.

Section 5

If a model is to be represented by algebraic equations then it is necessary to identify the *parameters* or variables appropriate to the situation being modelled and the relationship between them. Parameters differ from variables in that parameters have only *one value* within the context of the situation being modelled, whereas variables can take on *many different values* within the situation. *Symbols* can then be used to represent these parameters or variables and an algebraic equation written down. A *numerical value* can be found for the subject of the equation by *substituting* in numerical values for all the other parameters or variables in the equation.

Section 6.1

It is necessary to check that a *consistent set* of units is used for all the parameters and variables in the equation.

Section 6.2

A linear algebraic equation with only one 'unknown' can be *solved* by performing one or more *allowable operations* on it. These allowable operations are designed to keep the equation *balanced* while finding a numerical value for the 'unknown'.

Section 6.3

Since *inequalities* can be more difficult to manipulate than equalities, a problem is often stated in terms of an equality rather than an inequality. An inequality is represented graphically by a *region* on a graph.

Section 7

An alternative method of solving such equations is by using *function boxes*.

Appendix 1

ANSWERS TO SELF-ASSESSMENT QUESTIONS

SAQ 1

Since you were asked to prepare a similar model to the one in the text, you should have chosen one where the train travels at a constant speed all the way from passing the signpost to Paris.

Pairs of co-ordinates for the graph are: (0, 0), (1, 100), (2, 200), (3, 300), (4, 400), (5, 500). The first number in each pair represents the number of hours since passing the sign and the second number represents the number of kilometres that the train would travel.

The graph is drawn in Figure 44.

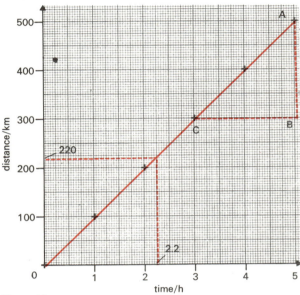

Figure 44

The gradient is 100. The lines BA and CB on Figure 44 are just one of the many pairs of lines you could have chosen in order to calculate the gradient.

Since Dijon is 220 km from the sign, Figure 44 shows a horizontal line from 220 to the line of the graph and a vertical line down to the horizontal axis. This vertical line intersects the horizontal axis at 2.2. The model therefore predicts that the train will arrive at Dijon about 2.2 hours after passing the sign.

SAQ 2

A line needs to be drawn up from $1\frac{3}{4}$ on the horizontal axis to the line of the graph and then across to the vertical axis, as shown in Figure 45.

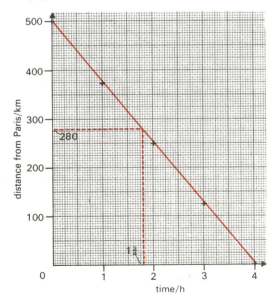

Figure 45

This line meets the vertical axis at 280, so Dijon is about 280 km from Paris.

SAQ 3

The modelling supposition you should have made is that the car travels at constant speed. This will lead to a straight-line graph. Table 12 shows the table of values and the co-ordinates.

Table 12 Data for the motorway journey

number of hours elapsed since joining the motorway	number of km travelled since joining the motorway	co-ordinates
0	0	(0, 0)
1	95	(1, 95)
2	190	(2, 190)
3	285	(3, 285)
4	380	(4, 380)
5	475	(5, 475)

The graph is shown in Figure 46.

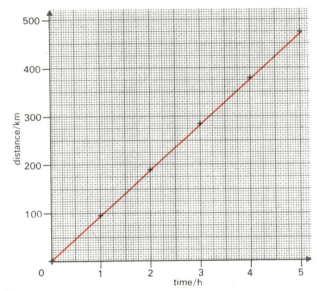

Figure 46

(a)   After 1.5 hours it has travelled 140 km, after 2.5 hours 240 km and after 4.4 hours 420 km.

(b)   It will have travelled 125 km after 1.3 hours and 200 km after 2.1 hours.

SAQ 4

The modelling supposition you should have made is that the boat travels at a constant speed. This will lead to a straight-line graph. Table 13 shows the table of values and the co-ordinates

Table 13 Data for the sea-crossing from England to France

number of hours elapsed since leaving England	number of km travelled since leaving England	co-ordinates
0	0	(0, 0)
$\frac{1}{2}$	10	$(\frac{1}{2}, 10)$
1	20	(1, 20)
$1\frac{1}{2}$	30	$(1\frac{1}{2}, 30)$
2	40	(2, 40)

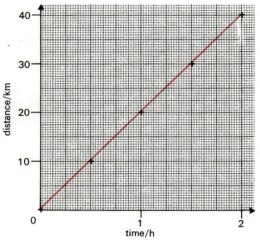

distance/km

time/h

Figure 47

The graph is shown in Figure 47. The gradient is 20.

(a)   The boat will be 15 km from England after $\frac{3}{4}$ hour, 25 km after $1\frac{1}{4}$ hours and 35 km after $1\frac{3}{4}$ hours.

(b)   After $\frac{1}{4}$ hour it has travelled 5 km, and after 1 hour 10 minutes it has travelled about 23 km.

You may have chosen to produce a table similar to Table 2, but you will probably have found that it does not really help you to answer the questions, since they were all posed in terms of distances and times from *England*.

SAQ 5

The graph is shown in Figure 48. He can take delivery after about $9\frac{1}{2}$ weeks.

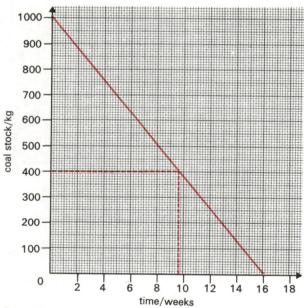

coal stock/kg

time/weeks

Figure 48

SAQ 6

Figure 49 shows how the two trains can be timetabled. The two trains should leave Station X about forty minutes apart, the slower train leaving first.

SAQ 7

(a)   The model is one of constant speed for the plane. Figure 50(a) shows the graph of the plane's motion if you have used a graph similar to that of Figure 6: Figure 50(b) shows the graph if you have used a graph similar to that of Figure 10. Either is acceptable.

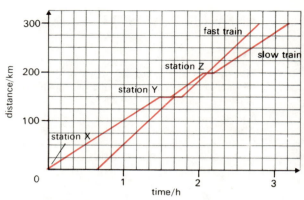

distance/km

time/h

Figure 49

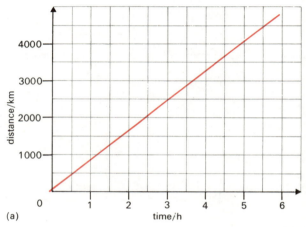

distance/km

(a)   time/h

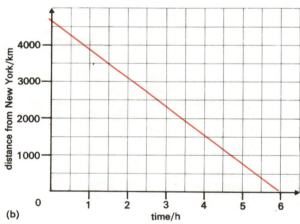

distance from New York/km

(b)   time/h

Figure 50

(b)   Six hours and forty minutes is $6\frac{2}{3}$ hours or 20/3 hours as an improper fraction. The average speed is given by:

$$\text{average speed} = 4800 \div \tfrac{20}{3}$$
$$= 4800 \times \tfrac{3}{20}$$
$$= 720.$$

So the average speed is 720 kilometres per hour.

Figures 51(a) and 51(b) show the graphs you might have drawn. Either is acceptable. Notice how the line representing the lower speed is less steep.

(c)   Since the average speed is reduced from 800 to 720 km per hour, it is reasonable to suppose that the headwind is blowing at 80 km per hour.

On the return journey, the wind will be a tailwind, so the plane will go faster than 800 kilometres per hour. With a tailwind, the average speed of the plane will be 800 + 80 = 880 kilometres per hour.

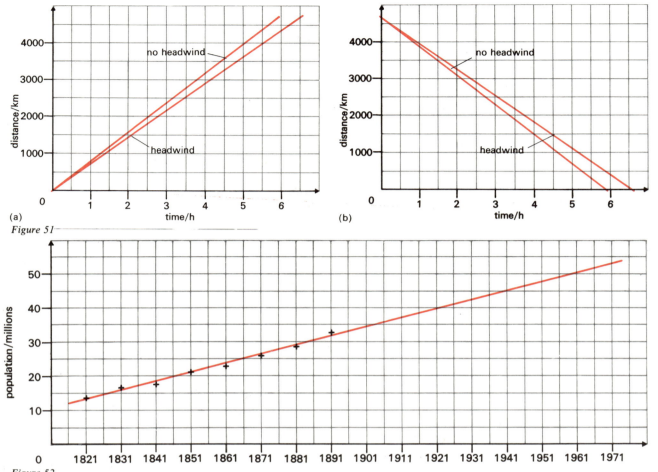

(a)

(b)

Figure 51

Figure 52

SAQ 8

Figure 52 shows the population figures plotted on a graph, with a model of the situation which assumes that the population was growing linearly and that any deviation of particular points from the 'average' line is just a small unimportant fluctuation. (You may have drawn a slightly different straight line from the one in Figure 52.)

By extrapolation, the graph predicts the populations, to nearest half million, to be as shown in Table 14.

Table 14

year	population
1901	34 500 000
1911	37 500 000
1921	40 000 000
1961	50 500 000
1971	53 000 000

The last two predictions are really not justified. 1971 is eighty years after the last point plotted on Figure 52, and Table 5 only spans seventy years anyway. The last two population figures are therefore being extrapolated over more than twice the range of the population figures given in the table—a very dangerous practice unless there is some theoretical reason for assuming that the population has risen in a linear fashion.

As you are aware, there were two wars in the twentieth century which influence the figures and may make even as early a prediction as that for 1921 incorrect. (You might like to look back to Figure 9 in Unit 1 which gave the actual figures for the population in the twentieth century.)

SAQ 9

The points are plotted in Figure 53, which represents a linear model of the supply of this commodity. (You may not have chosen exactly the

same line, which will influence your answers to questions (a), (b) and (c) slightly. Do not worry, provided your answers are about the right size they are acceptable.)

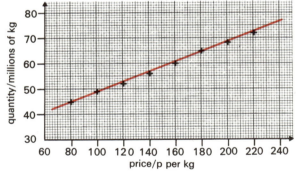

Figure 53

(a) At 70p per kg about 43 million kg would be supplied.

(b) At 130p per kg about 54 million kg would be supplied.

(c) At 240p per kg about 76 million kg would be supplied.

SAQ 10

The graph is shown in Figure 54, and the point of intersection shows that the police car should overtake the stolen car 8.3 minutes after the policeman spots it.

A model has been made because there is a modelling supposition that they both travel at a constant speed and that the speed they use is their maximum speed. There is a further supposition which you may have noted. This supposition is that they both reach their maximum speeds instantaneously—the stolen car as soon as its driver spots the police car and the police car as soon as it joins the motorway. In practice, this

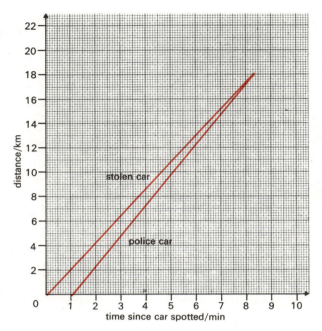

Figure 54

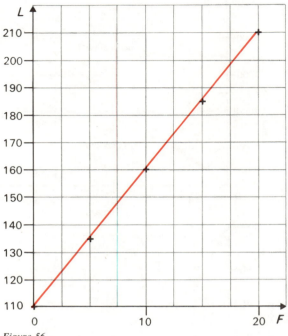

Figure 56

is not true. (Incidentally, my choice of kilometres per minute for the speeds of the cars was because the answer is most conveniently expressed in minutes, rather than hours.)

SAQ 11

Figure 55 shows the points plotted and lines drawn in to represent a model where both supply and demand are taken to be linear.

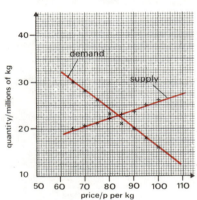

Figure 55

The point of intersection gives an approximate value for the equilibrium price for butter, which is about 83 pence per kilogram.

SAQ 12

(a) The independent variable is the force, because the length of the strip of rubber *depends on* the force pulling it.

(b) (i) When F is 10, L is 160; (ii) When F is 15, L is 185.

(c) The graph is shown in Figure 56: (i) F is 4; (ii) F is 14.

SAQ 13

The word equation is:

number of pence to seed the lawn

$$= \frac{\text{number of pence per kg of seed} \times \text{number of square metres to be seeded}}{\text{number of square metres covered by 1 kg of seed}}$$

SAQ 14

Given that: $P = 120$; $M = 100$; and $K = 20$.

$$C = 120 \times \frac{100}{20}$$

$$= 600$$

so the cost is 600p or £6·00.

SAQ 15

The parameters he needs to use to calculate 'the number of litres of paint to be used' are 'the number of square metres to be painted', 'the number of square metres covered per litre' and 'the number of coats of paint to be put on'.

SAQ 16

The relationship is as follows:

number of litres of paint to be used

$$= \frac{\text{number of coats of paint} \times \text{number of square metres}}{\text{number of square metres covered per litre}}$$

SAQ 17

(a) The extra parameter to enable the decorator to calculate 'the number of pounds the paint costs' is 'the number of pounds each litre of paint costs'. (If you have used pence it does not matter provided you have used pence for both costs.)

The relationship is:

number of pounds paint costs

$= $ number of pounds per litre $\times$ number of litres used

Using the relationship in SAQ 16, this can be rewritten as:

number of pounds paint costs

$$= \frac{\text{number of pounds per litre} \times \text{number of coats} \times \text{number of square metres to be painted}}{\text{number of square metres covered per litre}}$$

(b) You may have chosen different symbols to represent your parameters, but check that the algebraic equation you have written down corresponds to mine. I shall let:

C represent the number of pounds the paint costs

N represent the number of coats of paint to be put on

P represent the number of pounds cost per litre of paint

A stand for the number of square metres to be painted

L stand for the number of square metres covered per litre.

The algebraic equation is then

$$C = \frac{PNA}{L}$$

(c) Using the information given in the question

$P = 1.5$

$N = 2$

$A = 40$

$L = 20$

Substituting these values in the algebraic equation

$$C = \frac{1.5 \times 2 \times 40}{20}$$

$$= \frac{120}{20}$$

$$= 6.$$

The cost is therefore £6.

(d) In this case

$P = 1.2$

$N = 3$

$A = 60$

$L = 15$

Substituting in the algebraic equation

$$C = \frac{1.2 \times 3 \times 60}{15}$$

$$= \frac{216}{15}$$

$$= 14.4$$

The cost is therefore £14.40.

SAQ 18

(a) The parameters he should use to calculate 'the number of pounds the job costs' are 'the number of hours worked', 'the number of pounds charged per hour' and 'the number of pounds the paint costs'.

The relation is

number of pounds the job costs
= (number of hours worked × number of pounds charged per hour) + number of pounds the paint costs

(b) Again you may have chosen different symbols, but check your equation with mine for consistency. I shall let:

J stand for the number of pounds the job costs

C stand for the number of pounds the paint costs

H stand for the number of hours worked

B stand for the number of pounds charged per hour.

The algebraic equation is then

$$J = HB + C$$

(c) From the information given in the question

$H = 12$

$B = 2$

$C = 10$

Substituting these values in the algebraic equation

$$J = 12 \times 2 + 10$$

$$= 24 + 10$$

$$= 34$$

The job costs £34.

SAQ 19

(a) $C = 4$

(b) $A = 5$

(c) $P = 8$

(d) $U = 4$

(e) $B = -3$.

SAQ 20

(a) If $D = 250$ then

$$250 = 125 \times T$$

Divide both sides of the equation by 125.

$$\frac{250}{125} = \frac{125 \times T}{125}$$

Therefore $2 = T$ or $T = 2$.

(b) Similarly, when $D = 150$, $T = 1.2$.

(c) Similarly, when $D = 200$, $T = 1.6$.

(d) Similarly, when $D = 400$, $T = 3.2$.

SAQ 21

(a) If $D = 50$ then

$$50 = 100 \times T$$

Divide both sides of the equation by 100.

$$\frac{50}{100} = \frac{100 \times T}{100}$$

Therefore

$$0.5 = T \text{ or } T = 0.5.$$

(b) Similarly, when $D = 175$, $T = 1.75$.

(c) If $D = 100$ then

$$100 = 80 \times T$$

Divide both sides of the equation by 80.

$$\frac{100}{80} = \frac{80 \times T}{80}$$

Therefore $1.25 = T$ or $T = 1.25$.

(d) Similarly, when $D = 240$, $T = 3$.

SAQ 22

(a) (i) When

$$V = 5$$

$$5 = U - 20$$

Add 20 to both sides.

$$5 + 20 = U - 20 + 20$$

Therefore

$$25 = U \text{ or } U = 25.$$

(ii) Similarly, when $V = 30$, $U = 50$.

(b) (i) When $S = 9$

$$9 = 2U + 3$$

Subtract 3 from both sides.

$$9 - 3 = 2U + 3 - 3$$

Therefore $6 = 2U$

Divide both sides by 2.

$$\frac{6}{2} = \frac{2U}{2}$$

Therefore $3 = U$ or $U = 3$.

(ii) Similarly, when $S = 1$, $U = -1$.

(c) (i) When $S = 2$

$$2 = \frac{3 + 4V}{2}$$

Multiply both sides by 2.

$$2 \times 2 = 2 \times \frac{(3 + 4V)}{2}$$

Therefore $4 = 3 + 4V$

Subtract 3 from both sides.

$$4 - 3 = 3 + 4V - 3$$

Therefore $1 = 4V$.

Divide both sides by 4.

$$\frac{1}{4} = \frac{4V}{4}$$

Therefore $\frac{1}{4} = V$ or $V = \frac{1}{4}$

(Alternatively, $V = 0.25$.)

(ii) Similarly, when $S = 10$, $V = 4.25$ (or $V = 4\frac{1}{4}$).

(d) (i) When $L = 160$

$$160 = 110 + 5F$$

Subtract 110 from both sides.

$$160 - 110 = 110 + 5F - 110$$

Therefore $50 = 5F$

Divide both sides by 5.

$$\frac{50}{5} = \frac{5F}{5}$$

Therefore $10 = F$ or $F = 10$.

(ii) Similarly, when $L = 185$, $F = 15$.

(e) (i) When $D = 375$

$$375 = 500 - 125T$$

Subtract 500 from both sides.

$$375 - 500 = 500 - 125T - 500$$

Therefore $-125 = -125T$

Divide both sides by -125.

$$\frac{-125}{-125} = \frac{-125T}{-125}$$

Therefore $1 = T$ or $T = 1$.

(ii) Similarly, when $D = 0$, $T = 4$.

SAQ 23

(a) $2B + 9 = 7$

Subtract 9 from both sides of the equation.

$$2B + 9 - 9 = 7 - 9$$

Therefore $2B = -2$

Divide both sides of the equation by 2.

$$\frac{2B}{2} = \frac{-2}{2}$$

Therefore $B = -1$.

(b) $3Q - 11 = 22$

Add 11 to both sides of the equation.

$$3Q - 11 + 11 = 22 + 11$$

Therefore $3Q = 33$.

Divide both sides of the equation by 3.

$$\frac{3Q}{3} = \frac{33}{3}$$

Therefore $Q = 11$.

(c) $$\frac{P + 2}{5} = 6$$

Multiply both sides of the equation by 5.

$$5 \times \frac{(P + 2)}{5} = 5 \times 6$$

Therefore $P + 2 = 30$

Subtract 2 from both sides of the equation.

$$P + 2 - 2 = 30 - 2$$

Therefore $P = 28$.

SAQ 24

(a) $3A + 1 = 2A - 3$

Subtract $2A$ from both sides of the equation,

$$3A + 1 - 2A = 2A - 3 - 2A$$

Therefore $A + 1 = -3$

Subtract 1 from both sides of the equation.

$$A + 1 - 1 = -3 - 1$$

Therefore $A = -4$.

(b) $$\frac{2Y + 7}{3} = 3Y$$

Multiply both sides of the equation by 3.

$$3 \times \frac{(2Y + 7)}{3} = 3Y \times 3$$

Therefore $2Y + 7 = 9Y$

Subtract $2Y$ from both sides of the equation.

$$2Y + 7 - 2Y = 9Y - 2Y$$

Therefore $7 = 7Y$

Divide both sides of the equation by 7.

$$\frac{7}{7} = \frac{7Y}{7}$$

Therefore $1 = Y$ or $Y = 1$.

(c) $$Y + 1 = \frac{2Y + 3}{6}$$

Multiply both sides of the equation by 6.

$$6 \times (Y + 1) = 6 \times \frac{(2Y + 3)}{6}$$

Therefore $6 \times (Y + 1) = 2Y + 3$.

The left-hand side can be written $6Y + 6$ since the 6 outside the brackets multiplies both Y and 1, so

$$6Y + 6 = 2Y + 3.$$

Subtract $2Y$ from both sides of the equation.

$$6Y + 6 - 2Y = 2Y + 3 - 2Y$$

Therefore $4Y + 6 = 3$

Subtract 6 from both sides of the equation.

$$4Y + 6 - 6 = 3 - 6$$

Therefore

$$4Y = -3$$

Divide both sides of the equation by 4.

$$\frac{4Y}{4} = \frac{-3}{4}$$

Therefore $Y = -\frac{3}{4}$ (or $Y = -0.75$).

SAQ 25

The graph is shown in Figure 57.

(a) At 5 km from the town centre a producer of the second crop can afford a higher rent.

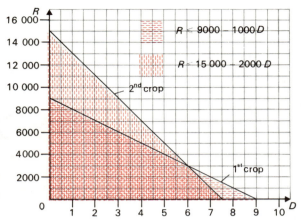

Figure 57

(b) At 8 km from the town centre a producer of the first crop can afford a higher rent. In fact, the producer of the second crop cannot afford *any* rent at this distance; it is no longer an economic proposition for him to grow the crop so far from the town centre.

The changeover occurs 6 km from the town centre.

SAQ 26

(a) $P = 2$. See Figure 58(a).

(b) $X = 6$. See Figure 58(b).

(c) $B = -1.5$ (or $-1\frac{1}{2}$). See Figure 58(c).

(d) $B = 21\frac{1}{3}$. See Figure 58(d).

(e) $P = 1\frac{1}{3}$. See Figure 58(e).

SAQ 27

(a) $X = -1$. See Figure 59(a).

(b) $X = 11.4$ (or $11\frac{2}{5}$). See Figure 59(b).

(c) $X = 10$. See Figure 59(c).

(d) $X = -7.5$ (or $-7\frac{1}{2}$). See Figure 59(d).

(e) $X = 7\frac{1}{3}$. See Figure 59(e).

SAQ 28

(a) $X = -2\frac{2}{5}$ (or -2.4).

(b) $X = 1\frac{1}{2}$ (or 1.5).

(c) $X = 1\frac{1}{2}$ (or 1.5).

(d) $X = -1$.

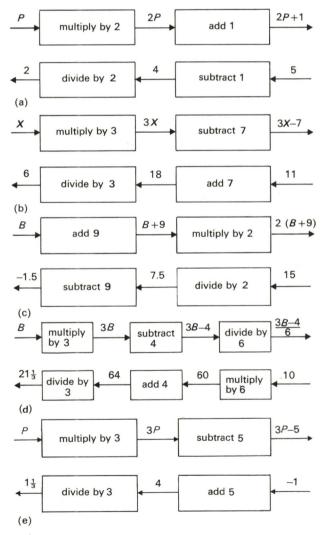

Figure 58

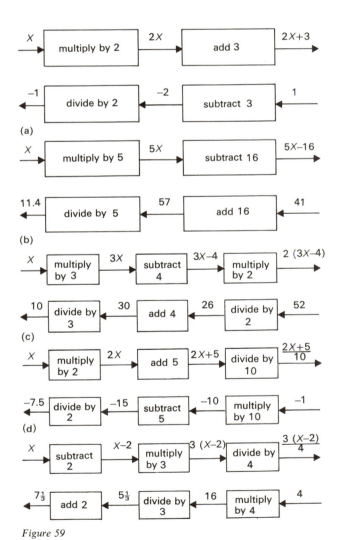

Figure 59

Acknowledgements

Grateful acknowledgement is made to the following for material used in this Unit:

Table 5 from *The British Economy Key Statistics 1900–1970*, London and Cambridge Economic Service; *Table 6* from B. R. Mitchell, *Abstract of British Historical Statistics*, Cambridge University Press.

3. Linear Models 2

CONTENTS

AIMS

The aims of this unit are:

1 To continue the work started in Unit 2 on deriving algebraic equations which represent models of given situations and to enable you to manipulate these algebraic equations.

2 To enable you to derive a pair of simultaneous equations which represent a model of a situation, to solve the equations and to relate the solution to the given situation.

3 To introduce you to the use of symbols to represent dimensioned quantities.

4 To give you a sufficient understanding of exponents and exponential curves to enable you to understand the way in which the scales on a slide rule are spaced and to use a slide rule to perform multiplication and division.

OBJECTIVES

When you have finished this unit you should be able to:

1 Distinguish between true and false statements concerning, or explain in your own words the meaning of, the following terms:

base

changing the subject of an equation

common factor

cube root

elimination

exponent

exponent form

exponential curve

index

intercept

irrational number

power

simultaneous equations

2 Carry out manipulations on an algebraic equation: such as, changing its subject (SAQs 5, 6, 7, 12), finding a common factor or manipulating brackets (SAQs 9, 10, 11) and manipulating algebraic fractions (SAQs 13, 14); and state why such manipulations may be useful in modelling by mathematics.

3 Solve an equation of the form $A/X = B$, where X is unknown (SAQ 8).

4 Make a model which leads to two linear simultaneous equations, solve these equations and relate the solution to the situation being modelled (SAQs 15, 16, 17, 18, 20, 21).

5 Check if a given equation is dimensionally balanced and state the advantages of using dimensioned quantities in equations (SAQ 22).

6 Give meaning to zero, negative and non-integer exponents and multiply and divide numbers containing exponents (SAQs 23, 24, 25).

7 Express a given number in exponent form and multiply and divide numbers in this form (SAQs 26, 27).

8 Use an exponential curve to find values of A^X, where X is a non-integer (SAQs 28, 29).

9 Use the slide rule to multiply and divide (*Slide Rule Book*).

STUDY GUIDE

Your work for this study week consists of reading Unit 3, *Linear Models 2*, watching the second television programme, *TV 2 Crossing by Numbers*, listening to three sides of discs, working with the slide rule and doing assignment material.

Unit 3, as its name suggests, is a continuation of the work started in Unit 2. It is important, therefore that you finish work on Unit 2 before you start to study this unit. Unit 3 contains some revision self-assessment questions to help you recall some of the more important points in Unit 2.

The television programme, *TV 2 Crossing by Numbers*, is related to all three units in this block and while it will help you to have read Unit 3 before you watch it, this is not essential. Side 2 of Disc 2 *Pedestrian delay* is associated with the television programme. Details are given in the broadcast notes for the programme.

Side 2 of Disc 1, *Using the slide rule (2)* illustrates how to use your slide rule to multiply and divide. You should listen to this disc as directed in Section 6.1 of this unit. You should also study Sections 4 and 5 of the *Slide Rule Book*, which contain examples and exercises on multiplication and division.

Side 1 of Disc 2, *Changing the subject of an equation*, is associated with Section 2. An example is given in the audiovision notes and the process is 'talked through' on the disc. Only if, as a result of answering the self-assessment questions on this topic in the unit, you feel quite confident about this process should you omit listening to this disc.

Full details of the assignment material associated with Unit 3 are given in the supplementary material.

What you have finished the unit you can use the objectives and summary to check that you know and can do what is expected of you before you go on to Unit 4.

1 INTRODUCTION

In the last section of Unit 2 you learnt how to derive simple algebraic equations which represent models of situations and how to substitute numbers into such equations. You also learnt how to solve equations with one 'unknown' in them. This unit begins by consolidating what you have already learnt and extending it. You will learn how to manipulate in various ways the algebraic equations that can be used to represent models of situations and how to solve equations which are slightly more complicated than those you have met so far. You will also learn about some of the modelling implications of these manipulations.

You should check your understanding of the material in Unit 2 by answering the two following self-assessment questions. Look back to Section 6 of Unit 2 if you have difficulty in answering either of them. The answer to SAQ 2 is used in Section 2, so it is particularly important that you try to answer it before starting Section 2.

SAQ 1 (revision)

Solve each of the following equations:

(a) $2A - 15 = -1$

(b) $3Z + 8 = 23$

(c) $3(B + 1) = 5$

(d) $\dfrac{2X - 3}{5} = 5$

(e) $2P + 2 = 5P - 4$.

SAQ 1

SAQ 2 (revision)

The population of a town is known at the present time and is known to be increasing steadily (that is, roughly the same number of people is added each year). An estimate of what the population will be at various times in the future is required.

(a) What parameters or variables should be used in order to derive an algebraic equation describing a model from which the future population can be estimated?

(b) Write down the relationship between them.

(c) Choose suitable symbols to represent them and hence write down an algebraic equation for the model of the situation.

(d) A town currently has a population of 22 000 and it is increasing at an average of 200 people per year. Give estimates for its population four, five and eight years from now. What do you estimate its population was two and three years ago?

(e) Why will your model only give an *estimate* of the future population?

SAQ 2

After consolidating your previous work, the unit goes on to discuss the algebraic solution of simultaneous equations. You have already met situations where two events can be modelled and, from the graphs representing these models, reached conclusions about where these events coincide. In this unit you will meet similar models of situations, but you will reach the conclusions about the coincidence of the events algebraically.

In Unit 2 you met briefly the idea that a symbol in an algebraic equation could be used either to represent a number or to represent a number *and* a unit, that is, a dimensioned quantity. Section 4 of this unit introduces a few simple cases where symbols are used to represent dimensioned quantities and acts as a foundation for further work on these lines later in the course.

The final section of the unit deals with the use of a slide rule for carrying out multiplication and division. This section is preceded by work on the topic of exponents, which is a necessary preliminary to the understanding of the slide rule.

2   MANIPULATION OF ALGEBRAIC EQUATIONS

2.1   A population model: changing the subject of an equation

In SAQ 2 you derived an algebraic equation representing a model which would enable you to estimate the future population of a town. I am going to use the answer to SAQ 2 as an example. Let:

N stand for the number of inhabitants at present.

A stand for the number of inhabitants added per year.

Y stand for the number of years from the present.

P stand for the number of inhabitants some years from the present.

Then the algebraic equation describing the model of the situation is

$$P = N + AY.$$

You will probably remember from Unit 2 that in algebraic equations a symbol may represent a variable (as in the equation $D = 125T$, where D and T were variables) or a parameter (as in the equation $C = PNA/L$, where P, N, A and L were all parameters). A variable can take on different values within one situation: in the case of $D = 125T$, the number of kilometres, D, varied from 0 to 500 as T, the number of hours, varied from 0 to 4. On the other hand, a parameter stays constant in any one situation, but may vary from one situation to another: in the case of $C = PNA/L$, the number of pounds cost per litre of paint, P, the number of coats applied, N, etc., all stayed constant in any one application.

In the equation $P = N + AY$ are P, A, N and Y parameters or variables?

A and N are parameters; P and Y are variables.

A represents the number of inhabitants added each year and so A stays unchanged in any one situation and is a parameter: similarly, N, which represents the number of inhabitants at present, stays unchanged in any one situation and so is a parameter. If the model is to be used to predict the future population, P, some number of years, Y, in the future, then both P and Y must vary within the situation and thus be variables. So in the case of the equation $P = N + AY$, some symbols represent variables and others parameters.

The algebraic equation $P = N + AY$ is a linear one; why?

Because a graph of $P = N + AY$, for given values of the parameters N and A, would be a straight line. P is increasing linearly as Y increases.

This linear algebraic equation has P as its subject. When numerical values for N, A and Y are substituted into it, a value for P can be found. As discussed in the answer to SAQ 2, this value of P will only be an *estimate* of the future number of inhabitants. (Turn to this answer to refresh your memory on this point if necessary.)

As an example, if Town A currently has a population of 59 000, growing at an average of 1200 per year, and an estimate of the population five years from now is required: $N = 59\,000$; $A = 1200$; and $Y = 5$. Substituting these numerical values into the equation

$$P = 59\,000 + 1200 \times 5$$
$$= 59\,000 + 6000$$
$$= 65\,000.$$

Therefore, an estimate of the population five years from now is 65 000 on this model.

Let me assume that the equation $P = N + AY$ has been used for Town A and has been found to give estimates for the future number of inhabitants that are sufficiently accurate for the town council's planning purposes. The council are therefore quite prepared to continue using the model, but now have a slightly different problem. They have been advised by their experts that the town's reservoirs can only adequately supply a population up to 70 000 and they want to know how long they have before the population reaches this figure and whether they should start to plan for a new reservoir immediately.

The equation they have used in the past is

$$P = N + AY$$

and because P is the subject, this equation is designed to calculate a value for P, given values for N, A and Y. For their current problem they want to find how long they have before the population reaches 70 000, that is, they want to find a value for Y, knowing the value for P.

Values for N and A will, of course, be known since N is the number of inhabitants at present and A is the number added each year, so in this situation, P, N and A have known values and Y is to be calculated.

You have already met one way of finding Y, which is to substitute in the numerical values for P, N and A and then solve the resulting equation.

> **SAQ 3** (revision)
>
> Find Y in the case where the current population of Town A is 59 000, growing at 1200 per year.

SAQ 3

This type of calculation of how many years in the future the population will reach a given value looks as if it may occur fairly often—when the council needs to check the adequacy of the hospital service, schools, sewage system, etc.—and so they may want the algebraic equation representing the model re-organized so that it is in the form where Y is the subject of the equation. Then it will be easy to use it to calculate a value for Y, given values for P, N and A. To make Y the subject of the equation it is necessary to go through a process similar in many ways to that of solving an equation, except that algebraic symbols (letters) are manipulated instead of numbers.

This process is called *solving an equation algebraically* or *changing the subject of an equation*. I shall refer to the process by the second of these two descriptions.

changing the subject of an equation

Since the algebraic expression is an equation, exactly the same allowable operations can be used to manipulate it as are used to solve the equations you met in Unit 2:

> Adding the same quantity to both sides of the equation.
>
> Subtracting the same quantity from both sides of the equation.
>
> Multiplying both sides of the equation by the same quantity.
>
> Dividing both sides of the equation by the same (non-zero) quantity.

Since Y is to be made the subject of the equation

$$P = N + AY$$

the equation should be manipulated so that Y stands alone on one side of the equals sign and does not occur at all on the other side.

To form $N + AY$, Y must have been multiplied by A and then N added to this product. Thus the first step in getting Y to stand alone is to subtract N from both sides of the equation.

$$P - N = N + AY - N$$

Therefore

$$P - N = AY.$$

The next step is to divide both sides of the equation by A.

$$\frac{P - N}{A} = \frac{AY}{A}$$

Therefore

$$\frac{P - N}{A} = Y.$$

Y now stands alone on one side of the equation and so is the new subject of the equation.

I have had to divide both sides of the equation by the quantity A and this is only an allowable operation if A is not zero. Since A stands for the number of inhabitants added each year, it is not zero for Town A and there is no problem.

Using this new form of the equation representing the model of the town's population

$$Y = \frac{P - N}{A}$$

values of Y can be calculated quickly for various values of the future number of inhabitants, P.

Before going any further I want to discuss what I have just done from a modelling viewpoint. I took an equation

$$P = N + AY$$

which was known to be an equation representing a useful model for predicting the future population of Town A, and I manipulated it mathematically into the form

$$Y = \frac{P - N}{A}.$$

I took a mathematical expression which represented a model of a situation and which therefore had strong ties with the situation and I just manipulated it according to mathematical rules as if it had had no such ties at all! In modelling by mathematics, it is important for it to be possible to do just this, because the modeller wants to find out something about a situation which is not immediately obvious from just thinking about the situation. He does this by: (1) making a model of the situation; (2) describing the model by a mathematical expression which relates the variables or parameters to each other in such a way that the expression can be readily handled mathematically and is also a close enough approximation to the situation to satisfy his purposes; and (3) manipulating this expression so that the required information is obtained. In other words, he goes through the modelling cycle illustrated in Figure 12 of Unit 1. If, after the manipulation has been correctly carried out, the result appears unlikely or even nonsensical, then it is highly probable that the original model and its associated mathematical expressions were not the right ones to use in those circumstances.

SAQ 4
SAQ 4

At a time when the population of Town A is 65 000 and still growing at 1200 per year, it is estimated that, to maintain the same level of service: the water resources can serve a town of up to 70 000 inhabitants; and the current staffing levels at the Town Hall can adequately cope with rate demands for up to 21 000 households, the average number of inhabitants per household being 3.5.

After how long will adjustments have to be made to:

(a) The water resources.

(b) The number of staff at the Town Hall dealing with rates?

SAQ 5

In this question do not worry about what the symbols in the equation mean; it is designed to give you practice in changing the subject of algebraic equations.

(a) Make T the subject of $D = 125T$

(b) Make F the subject of $L = 110 + 5F$

(c) Make U the subject of $V = U + AT$

(d) Make T the subject of $V = U + AT$ (A is non-zero)

(e) Make N the subject of $P = N + AY$.

SAQ 6

In SAQs 15–17 of Unit 2 you devised a model and wrote down an associated algebraic equation which could be used to calculate the cost of painting the walls of a room. The algebraic equation was

$$C = \frac{PNA}{L}$$

where C represents the number of pounds the paint costs; P the number of pounds each litre of paint costs; N the number of coats of paint to be put on; A the number of square metres to be painted; L the number of square metres covered per litre.

(a) A home decorator using this model to calculate his costs does not want to spend more than a fixed amount on repainting a room and is wondering what is the maximum he can spend per litre of paint. What should be the new subject of the equation

$$C = \frac{PNA}{L}$$

in order for him to use the equation to perform the calculation he needs?

(b) Change the subject of the equation as required.

(c) His room has an area of forty-eight square metres and needs two coats of paint. Assume that one litre of paint covers sixteen square metres. He has £9 to spend. What is the maximum he can spend on one litre of paint?

In SAQ 6 you have just been manipulating the equation you first met in Unit 2 for calculating the cost of repainting a room and so made P the subject of the equation

$$C = \frac{PNA}{L}.$$

Suppose you were given the equation

$$C = \frac{PNA}{L}$$

and asked to make L the subject. This is slightly more difficult because L appears in the denominator of one side of the equation, but it can be done by making the first allowable operation one that puts L into the numerator of the other side of the equation.

This can be done if both sides of the equation are multiplied by L.

$$CL = \frac{PNA}{L} \times L$$

12

Thus
$$CL = PNA.$$
Both sides are now divided by C.
$$\frac{CL}{C} = \frac{PNA}{C}$$
and so
$$L = \frac{PNA}{C}$$
(assuming C is not equal to zero).

SAQ 7

SAQ 7

Make K the subject of
$$C = \frac{PM}{K}$$
(assuming C is not equal to zero).

A very similar process can be used to solve equations which are slightly more complicated than those you met in Unit 2. For example, to solve
$$\frac{2}{X} = 3$$
the first step is to get X into the numerator on one side by multiplying both sides of the equation by X.
$$\frac{2}{X} \times X = 3X$$
Therefore
$$2 = 3X.$$
The solution is $X = 2/3$.

SAQ 8

SAQ 8

Solve each of the following equations:

(a) $\dfrac{1}{X} = 4$

(b) $\dfrac{5}{X} = 7$

(c) $\dfrac{9}{2X} = \dfrac{1}{6}$

(d) $\dfrac{3}{5X} = 1$

2.2 A model for redecorating: the use of brackets

I am going to continue using the equation
$$C = \frac{PNA}{L}$$
which is an equation representing a model used to calculate the cost of painting the walls of a room. In this equation, the parameter A, the number of square metres to be covered, can only be given a numerical value by performing a calculation to evaluate it for a given room. This calculation will involve numerical values for the length, breadth and height

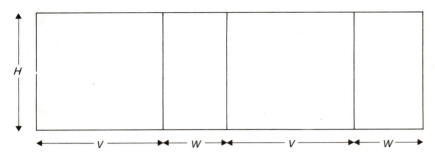

Figure 1

of the room. Figure 1 shows the walls of the room 'unwrapped' and laid out flat. The area enclosed by any rectangle can be found by multiplying together the length of its longer side and the length of its shorter side. (You probably know this as 'area of rectangle = length × width'.) From Figure 1 you can see that an expression for *A*, neglecting the doors and windows, is

$$A = HV + HW + HV + HW$$

where *V* stands for the number of metres in the length of the room; *W* stands for the number of metres in the width of the room; *H* stands for the number of metres in the height of the room.

Now *HV* added to another *HV* makes *2HV* and *HW* added to another *HW* makes *2HW*, so

$$A = 2HV + 2HW.$$

Both of the terms on the right-hand side of this equation have 2 and *H* occurring in them and in both terms they are multiplying something— multiplying *V* in *2HV* and *W* in *2HW*. Thus, *2H* is said to be a *common factor* of these two terms. When a common factor occurs like this, it can be convenient to rewrite the equation in the following way

$$A = 2H(V + W).$$

common factor

Whenever such common factors occur outside a bracket, whatever appears outside the bracket multiplies *each* term in the bracket. That is, *2H* must multiply *both V* and *W* so that the height is still multiplying both the length and the width. With this convention

$$2H(V + W) \qquad \text{and} \qquad 2HV + 2HW$$

are equal to each other, as they should be if I am going to rewrite

$$A = 2HV + 2HW$$

as

$$A = 2H(V + W).$$

At this stage I need to point out the rules for the order in which a mixture of addition, subtraction, multiplication and division operations should be carried out. The example following the rules should make them clear. They are, in the order in which they should be carried out:

1   Any operation enclosed in brackets is carried out first, and where more than one pair of brackets are used, the operation in the innermost brackets is carried out first.

2   All multiplications and divisions are carried out before any addition and subtraction.

3   Additions and subtractions are carried out in order, working from left to right.

As an example, I shall evaluate the following:

$$[3(2 + 5 - 4) - 6] \div 10 + 2$$

Rule 1 states that the operation in the innermost brackets is carried out first.

To carry out the operation in the inner brackets, Rule 3 is applied: $2 + 5 - 4$ is $7 - 4$ which is 3, so the expression becomes

$$[3 \times 3 - 6] \div 10 + 2.$$

We still have brackets in this expression so Rule 1 must be applied, but there are two operations to be carried out, multiplication and subtraction. Rule 2 states that multiplication is carried out before subtraction: hence

$$[9 - 6] \div 10 + 2.$$

Rule 1 is again applied and this expression becomes

$$3 \div 10 + 2.$$

There are no more brackets and Rule 2 states that the division is carried out before the addition.

$$0.3 + 2.$$

The result is therefore 2.3.

I have chosen to illustrate the rules with a numerical example, but identical rules apply to algebraic expressions which include a mixture of operations.

In the case of

$$A = 2H(V + W)$$

the rules show that when this expression is being evaluated, numerical values for V and W should be added together before being multiplied by the numerical value of $2H$, so that $2H(V + W)$ means two multiplied by H multiplied by the result of adding V to W.

In rewriting

$$A = HV + HW + HV + HW$$

as

$$A = 2H(V + W)$$

you may feel that the equation is no longer obviously closely connected to the real-life situation. In the first way of writing the expression for A, each term— that is, HV, HW, HV and HW—is obviously the area of one of the four walls, while in

$$A = 2H(V + W)$$

it is not quite so clear just how $2H(V + W)$ is related to the area of the four walls. However, such a re-arrangement is valuable mathematically for two reasons. One is that it often makes an expression easier to evaluate. For instance, in evaluating $2H(V + W)$ there is only one addition followed by two multiplications whereas in $2HV + 2HW$ there are four multiplications; two to evaluate $2HV$ and two to evaluate $2HW$, and then one addition. The other reason is that in many cases it makes the expression more compact and easier to manipulate. The fact that in carrying out such manipulations the equation may seem to lose some of its connections with the situation being modelled is something which may happen in modelling by mathematics. It is worthwhile if it simplifies the mathematics and, if the manipulation has been done correctly, it will be possible to relate the results to the situation quite readily. For instance, $2H(V + W)$ can be illustrated as in Figure 2 and you can see that, by laying out the four areas in a different way from that in Figure 1, I have been able to relate $2H(V + W)$ to the area of the four walls.

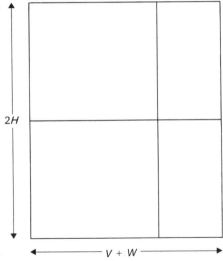

Figure 2

I can substitute this expression $A = 2H(V + W)$ into

$$C = \frac{PNA}{L}$$

giving

$$C = \frac{PN2H(V + W)}{L} = \frac{2PNH(V + W)}{L}.$$

In a mixture of letters and numbers, it is usual to write the number first.

15

In the numerator of the right-hand side, $2PNH(V + W)$, does $2PNH$ multiply both V and W?

Yes.

You may be wondering how you would treat an equation which included brackets, like this one, if you wanted to change the subject of the equation. Suppose, for instance, you wanted to make P the subject of

$$C = \frac{2PNH(V + W)}{L}.$$

The method is to treat $(V + W)$ as a single quantity as far as possible while performing allowable operations on the equation, as follows.

Multiply both sides of the equation by L.

$$CL = 2PNH(V + W)$$

Divide both sides by $(V + W)$.

$$\frac{CL}{(V + W)} = 2PNH$$

Divide both sides by 2.

$$\frac{CL}{2(V + W)} = PNH$$

Divide both sides by N.

$$\frac{CL}{2N(V + W)} = PH$$

Divide both sides by H.

$$\frac{CL}{2NH(V + W)} = P.$$

I have divided both sides of the equation by the quantities N, H and $(V + W)$, so it is necessary to check that none of these is zero, otherwise I have not performed allowable operations. Remembering what N, H, V and W stand for, you can see that they are not equal to zero, so there is no problem in dividing by any of N, H and $(V + W)$.

You may have noticed that, when dividing by N and again when dividing by H, I have arranged that the number 2 comes first and the bracketed term $(V + W)$ comes last. As well as it being usual to put the number first, any term in brackets is usually put last and this is what I did here.

SAQ 9

Suppose that a professional decorator is asked to paint a room. Added to the cost of the material, C, will be the cost of the labour. A model for the cost of the labour can be represented as follows:

$$B = \frac{DNA}{M}$$

where B stands for the number of pounds the labour costs; D stands for the number of pounds the decorator charges per hour; M stands for the number of square metres covered per hour; and N and A have the meaning they have had throughout this decorating model, that is, N stands for the number of coats to be put on and A the number of square metres to be covered.

Then if the number of pounds the whole job costs is represented by J

$$J = C + B$$

and substituting the expressions derived for C and B

$$J = \frac{PNA}{L} + \frac{DNA}{M}.$$

Find the common factor in the right-hand side of this equation and hence rewrite the right-hand side using brackets.

SAQ 10

In this question do not worry about what the symbols in the equation mean; it is designed to give you practice in spotting common factors. In each of the following, find the common factor and rewrite the equation using brackets.

(a) $S = \dfrac{UT}{2} + \dfrac{VT}{2}$

(b) $S = UT + \frac{1}{2}AT^2$ (Hint: remember that T^2 is $T \times T$)

(c) $Y = 3X^2 - 2X$

SAQ 11

Once again, do not worry what the symbols mean. In each of the following, write down what the expression would be if brackets were not used. This manipulation is sometimes referred to as 'removing the brackets' or 'multiplying out the brackets'.

(a) $Y = X(3X - 1)$

(b) $A = 2B(5 - 7B)$

(c) $D = 25(3T - 2)$

(d) $I = M(V - U)$.

SAQ 12

Once again do not worry about what the symbols in the equations mean.

(a) Make T the subject of the equation

$$S = \frac{(U + V)T}{2}$$

(the sum $(U + V)$ is non-zero).

(b) Make V the subject of the equation

$$S = \frac{(U + V)T}{2}$$

(T is non-zero).

In SAQ 9 you found that

$$J = \frac{PNA}{L} + \frac{DNA}{M}$$

can be rewritten as

$$J = NA\left(\frac{P}{L} + \frac{D}{M}\right)$$

Both P/L and D/M are algebraic fractions, and just as there are rules for adding numerical fractions (such as $\frac{1}{4} + \frac{1}{6}$), so there are rules for adding algebraic fractions, like $P/L + D/M$. In adding $\frac{1}{4} + \frac{1}{6}$, the first step is to find the same denominator for both fractions (a process referred to as 'finding a common denominator'), in this case 12. The next step is to adjust the numerators of both $\frac{1}{4}$ and $\frac{1}{6}$ to fit with their new denominator of 12. Thus $\frac{1}{4}$ becomes $\frac{3}{12}$ and $\frac{1}{6}$ becomes $\frac{2}{12}$. The final step is to add the numerators.

$$\frac{3 + 2}{12} = \frac{5}{12}$$

The steps are just the same in adding an algebraic fraction: (1) find a common denominator; (2) adjust the numerators of each fraction to fit this new denominator; (3) add the numerators.

In the case of

$$\frac{P}{L} + \frac{D}{M}$$

17

(1) a common denominator is LM.

(2) Adjusting the numerators and denominators.

$$\frac{P}{L} \text{ becomes } \frac{PM}{LM}$$

and

$$\frac{D}{M} \text{ becomes } \frac{LD}{LM}$$

so that

$$\frac{P}{L} + \frac{D}{M} = \frac{PM}{LM} + \frac{LD}{LM}$$

(3) Adding the numerators.

$$\frac{P}{L} + \frac{D}{M} = \frac{PM}{LM} + \frac{LD}{LM}$$
$$= \frac{PM + LD}{LM}$$

Using the fact that

$$\frac{P}{L} + \frac{D}{M} = \frac{PM + LD}{LM}$$

the equation

$$J = NA \left(\frac{P}{L} + \frac{D}{M} \right)$$

can be rewritten as

$$J = NA \frac{(PM + LD)}{LM}$$
$$= \frac{NA}{LM} (PM + LD)$$

The reorganization of the terms within the bracket in an equation, using the rule for manipulating algebraic fractions, is of particular value when the subject of the equation is changed. For instance, suppose A was to be made the subject of the original equation

$$J = NA \left(\frac{P}{L} + \frac{D}{M} \right)$$

The result would be

$$A = \frac{J}{N(P/L + D/M)}$$

where neither N nor $(P/L + D/M)$ is zero. This last equation is very cumbersome. It is not generally considered very 'neat' to leave fractions in the denominator in this way.

If, on the other hand, A was to be made the subject of

$$J = \frac{NA}{LM} (PM + LD)$$

then

$$A = \frac{JLM}{N(PM + LD)}$$

where neither N nor $(PM + LD)$ is zero. There are no fractions in the denominator of this last equation.

Subtraction of algebraic fractions follows an identical rule to addition, except that the numerators are subtracted, rather than added.

SAQ 13

Rearrange each of the following so they have a common denominator:

(a) $\dfrac{1}{X} + \dfrac{2}{Y}$

(b) $\dfrac{1}{A} - \dfrac{B}{C}$

(c) $\dfrac{P}{QR} - \dfrac{S}{R}$

SAQ 14

Rearrange the following so the fractions in the bracket have a common denominator

$$P = L\left(\frac{C}{NA} - \frac{D}{M}\right)$$

2.3 Summary

In this section you have learnt some useful algebraic techniques.

The first technique was changing the subject of an algebraic equation. This can be done by using the four allowable operations you met in Unit 2.

The next technique was that of finding a common factor in a group of terms and using brackets to express such a group of terms in a more compact form.

The third technique was that of expressing algebraic fractions in a form with a common denominator by applying similar rules to those used in adding and subtracting numerical fractions.

These techniques are useful in manipulating the algebraic expressions obtained in modelling situations and, provided the model has been correctly chosen for its given purpose, should yield results which are relevant to the situation being modelled.

Study comment

Side 1 of Disc 2, *Changing the subject of an equation* **is related to the work in this section, and you should listen to this disc either now or in the near future, unless, as a result of answering the self-assessment questions in Section 2, you feel fully confident of your ability to perform this manipulation.**

3 MODELS REPRESENTED BY PAIRS OF LINEAR ALGEBRAIC EQUATIONS

In Unit 2, I discussed some of the ways in which people who prepare timetables can ensure that fast and slow trains can use the same track as efficiently as possible. Such methods depend upon trains running to time. I want to look at an example of a case where a slow train is delayed and try to find an algebraic method of investigating whether a faster train following it along the line is held up as a consequence.

Figure 3 shows the situation. Stations A, B and C are spaced along a stretch of railway with only one track in each direction except at the stations. Normally, a slow train, which stops at all three stations, reaches Station C where a faster train, which only stops at Station A, passes it while it is stationary. On this particular occasion, the slower train is late and leaves Station B at exactly the same time as the faster train leaves Station A. The slower train's speed is 80 km per hour, the faster train's speed is 120 km per hour. Will the faster train have to be delayed?

Figure 3

To answer this question I need to make a model of the situation and then write down algebraic equations representing the model.

The suppositions I shall make in the model are:

1   Both trains reach their normal speeds immediately and maintain them throughout the journey (and, in the case of the slower train, that it continues at this speed until it stops).

2   The lengths of the trains are unimportant and each train can be treated as if its whole length was the same distance from Station A.

3   The faster train is not impeded by meeting warning amber signals that restrict its speed when it is still as much as two or three kilometres behind the slower train.

I shall measure distances from Station A, and I shall let: D_F represent the number of kilometres the faster train is from Station A: D_S represent the number of kilometres the slower train is from Station A; and T represent the number of hours that have elapsed since both trains moved off from their respective stations.

Notice that as the speeds are given in kilometres per *hour* it is appropriate to let T represent a number of hours so that the units are consistent.

I have used two subscripts in the above symbols. The subscript S is used in D_S and the subscript F in D_F. This use of subscripts is quite common when symbols are used to represent variables. Since both the number of kilometres travelled by the faster train and the number travelled by the slower train are distances, it is convenient to use the letter D for both, but as these two distances will be different (except at the moment when one train passes the other) the same letter can be used only if distinguishing subscripts are used— F for the faster train, S for the slower. The symbols D_F and D_S represent *different* distances, but they each represent a distance.

I shall first try to find an equation which can be used to represent the model of the faster train's motion. The variables to be related are D_F, the number of kilometres this train is from Station A, and T, the number of hours that have elapsed since it moved off. The speed at which the train is travelling will clearly come into the relationship.

The relationship can be deduced as follows. When one hour has elapsed, the faster train has moved 1×120 km from Station A; when two hours have elapsed, the faster train has moved 2×120 km from Station A, etc.

When T hours have elapsed, how many kilometres will the train have moved from Station A?

$T \times 120$, which can be written as $120T$.

But D_F represents the number of kilometres the faster train has moved from Station A; therefore,

$$D_F = 120T.$$

This equation is only true after the train leaves Station A, so it should have the qualification $T \geq 0$ alongside, where $\geq$ means 'is greater than or equal to'. Therefore, the equation representing the model of the faster train's motion is

$$D_F = 120T \qquad (T \geq 0).$$

Having derived an equation for the model of the faster train's motion, the next step is to derive an equation for the model of the slower train's motion.

What are the variables to be related in this case?

D_S, the number of kilometres the slower train is from Station A, and T the number of hours elapsed since the two trains started.

What else will come into the relationship?

The speed this train is travelling and the distance it starts from Station A, that is, how far Station B is from Station A.

I can deduce the relationship by temporarily ignoring the fact that the train stops at Station C, but putting it in as a limit on the validity of the equation afterwards: when one hour has elapsed the train has moved 1×80 km; when two hours have elapsed the train has moved 2×80 km; and so on.

When T hours have elapsed, how many kilometres has the train moved from Station B?

$80T$.

The number of kilometres this train has moved after T hours is not the same as the number of kilometres this train is from Station A (represented by D_S).

Why?

Because it starts 30 km along the line at Station B.

This means that, in order to find D_S, 30 must be added to the number of kilometres this train has moved.

The relationship is therefore

$$D_S = 30 + 80T.$$

This equation should include two restrictions. One is that $T \geq 0$. The other should limit it to not travelling beyond Station C. Since Station C is 60 km from Station B, and since the train travels at 80 km per hour, this restriction must be that $T \leq \frac{3}{4}$. These two restrictions can be combined into the statement $0 \leq T \leq \frac{3}{4}$, so

$$D_S = 30 + 80T \qquad (0 \leq T \leq \tfrac{3}{4}).$$

There are now two equations for the model of the situation, one for the faster train's motion and the other for the slower train's.

$$D_F = 120T \qquad (T \geq 0)$$
$$D_S = 30 + 80T \qquad (0 \leq T \leq \tfrac{3}{4})$$

When you drew graphs in Unit 2, it was easy to spot the point where the two events coincided: it was where the lines crossed. How can the 'coincidence' of the *equations* be found?

We are trying to find when the faster train passes the slower one in order to find out if it occurs when the slower train is at Station C. At the moment of overtaking, D_S will be equal to D_F, because both trains must be the *same* number of kilometres from Station A at that instant. Suppose that this distance from Station A is $\bar{D}$ kilometres (read as 'D bar') and that the overtaking occurs $\bar{T}$ hours (read as 'T bar') after they move off.

Do the symbols $\bar{D}$ and $\bar{T}$ represent variables or parameters?

At the time of overtaking $T = \bar{T}$ and $D_F = D_S = \bar{D}$. Therefore, for the faster train

$$\bar{D} = 120\bar{T}$$

and for the slower train

$$\bar{D} = 30 + 80\bar{T}.$$

The values of $\bar{D}$ and $\bar{T}$ are specific to the situation and do not vary within the situation. They are therefore parameters. That is why I have given them different (but related) symbols from D_F, D_S and T, which represent variables.

I now have a pair of equations *both* of which are true for $\bar{D}$ and $\bar{T}$: that is, I have a pair of *simultaneous equations*

simultaneous equations

$$\bar{D} = 120\bar{T}$$
$$\bar{D} = 30 + 80\bar{T}$$

If I can solve these two equations I can find out whether the faster train is impeded by the slower train (to the accuracy of my model).

To solve these two equations I need only use the allowable operations you have already met, though in a rather cunning way. The aim of the operations will be to *eliminate* one of the two 'unknowns', $\bar{D}$ or $\bar{T}$, and derive just one equation which, as it contains only one 'unknown' will be solvable. Before I do this for the pair of equations above I shall illustrate the method with two other pairs of equations in the two following examples.

elimination

Example 1
Solve the following pair of simultaneous equations: that is, find A and B.

$$2A = 7 + B \qquad (1)$$
$$A = 5 - B \qquad (2)$$

You already know that it is an allowable operation to add the same quantity to both sides of an equation, so I shall take equation (2) and add the same quantity to both sides of it. The quantity I shall add is $2A$ to the *left*-hand side and $7 + B$ to the *right*-hand side. Notice that I am adding the same quantity to both sides of equation (2) because equation (1) tells me that $2A$ *equals* $7 + B$. This operation is in fact called 'adding the equations'.

The result of this addition is

$$A + 2A = 5 - B + 7 + B$$

which is $3A = 12$.

Notice that the result of the addition is to eliminate one of the 'unknowns', B in this case, and to leave one equation

$$3A = 12.$$

This can be solved by dividing both sides by three.

$$A = 4.$$

I can now substitute $A = 4$ into either equation (1) or equation (2) and find B. Suppose I use equation (1)

$$2A = 7 + B$$

By substituting 4 instead of A in this equation I obtain

$$8 = 7 + B$$

so that $B = 1$.

So the answer is $A = 4$, $B = 1$.

I found B by substituting in *equation* (*1*). As a check, do the values $B = 1$ and $A = 4$ fit with *equation* (2)? Equation (2) says

$$A = 5 - B$$

and 4 does equal $5 - 1$. Therefore, the solution fits this equation.

Example 2
As a second example, solve the pair of simultaneous equations

$$3X = 7 + Y \tag{1}$$
$$2X = 2 + Y \tag{2}$$

Try adding the same quantity to both sides of the equation as in my previous example. Does it work in this case?

'Adding the equations' this time gives the result

$$5X = 9 + 2Y$$

Although you now have only one equation, you have not eliminated either of the 'unknowns', X or Y.

Can you suggest how to obtain one equation with one of the 'unknowns' eliminated?

This time, Y can be eliminated by *subtracting* the same quantity from both sides of equation (2). The quantity to be subtracted is $3X$ from the *left*-hand side and $7 + Y$ from the *right*-hand side. Again, I am subtracting the same quantity from both sides of equation (2), since equation (1) tells me that $3X$ equals $7 + Y$.

If I carry out this process, called 'subtracting the two equations', I obtain

$$2X - 3X = 2 + Y - (7 + Y)$$
$$= 2 + Y - 7 - Y$$

so

$$-X = -5$$

which means $X = 5$.

Substituting this value for X into equation (1)

$$15 = 7 + Y$$

and

$$Y = 8.$$

So the answer is $X = 5$, $Y = 8$.

As a check, equation (2) states that $2X = 2 + Y$, and 2×5 does equal $2 + 8$.

Solve the following pairs of simultaneous equations and check your answers:

(a) $X = 2Y - 3$ (1)

      $3X = 2Y + 5$ (2)

(b) $2A = 5B + 2$ (1)

      $2A = 3B - 2$ (2)

Returning to the problem of the two trains

$$\bar{D} = 120\bar{T} \qquad\qquad\qquad\qquad (1)$$

$$\bar{D} = 30 + 80\bar{T} \qquad\qquad\qquad (2)$$

These simultaneous equations can be solved by subtracting the equations.*
I shall subtract $\bar{D}$ from the *left*-hand side of equation (1) and $30 + 80\bar{T}$
from the *right*-hand side of equation (1).

$$\bar{D} - \bar{D} = 120\bar{T} - (30 + 80\bar{T})$$
$$= 120\bar{T} - 30 - 80\bar{T}$$

so

$$0 = 40\bar{T} - 30.$$

Adding 30 to both sides.

$$30 = 40\bar{T}$$

Dividing both sides by 40.

$$\frac{30}{40} = \bar{T}$$

So $\bar{T} = \tfrac{3}{4}.$

$\bar{T} = \tfrac{3}{4}$ is right on the limit of validity of equation (2), but it is not outside the
limits, so $\bar{T} = \tfrac{3}{4}$ can be substituted into either equation (1) or equation (2) to
give a value for $\bar{D}$. Substituting into equation (1)

$$\bar{D} = 120 \times \tfrac{3}{4}$$

so

$$\bar{D} = 90.$$

Equation (2) shows $\bar{D} = 30 + 80\bar{T}$ and 90 does equal $30 + 80 \times \tfrac{3}{4}$.

The overtaking on this model would take place 90 km from Station A and,
looking back to the original problem, Station C is 90 km from Station A.

This looks fine, the slow train would be at Station C, thus allowing the fast
train to pass safely. But let us look more closely at the situation being
modelled, at the modelling suppositions and at the results. The following two
points emerge:

1   The results show that the slow train will *only just* have reached Station C.

2   The model took no account of points like how quickly the trains
accelerate or how long they are and also it took no account of the fact
that the faster train would be alerted if the slower train was only two to
three kilometres ahead of it on the track.

Would you say that the faster train is likely to be delayed by the slower train on
this day when the slower train is late?

---

* You may have spotted that since $\bar{D} = 120\bar{T}$ and $\bar{D} = 30 + 80\bar{T}$ then it is possible to write $120\bar{T}$
$= 30 + 80\bar{T}$. This is a correct method of proceeding and will give the same answer, but it does
not illustrate the more general method of solving simultaneous equations, which is what I am
trying to do here.

My conclusion is that it would be. A situation where the slower train is just moving off the track to stop at Station C while the faster train pounds down on its tail is not one that should happen on the railways—trains are kept at least 1–2 km apart if they are using the same track.

So far, you have met some examples where it was possible to solve the simultaneous equations by adding one equation to the other or subtracting one equation from the other. Sometimes it is necessary to manipulate one of the equations first, as in the next example.

Example 3

Solve the equations

$$2X = 3 - 3Y \tag{1}$$
$$X = 4 + Y \tag{2}$$

Neither adding one equation to the other nor subtracting one from the other will eliminate either 'unknown'. This is because X appears in one equation while $2X$ appears in the other and Y appears in one, but $3Y$ in the other. The equations can be solved provided both sides of one equation are first multiplied by the same quantity. I shall try to eliminate Y, and I can do this if I first multiply both sides of equation (2) by 3.

$$3X = 12 + 3Y$$

I can now add $2X$ to the *left*-hand side and $3 - 3Y$ to the *right*-hand side, since $2X = 3 - 3Y$, from equation (1).

$$3X + 2X = 12 + 3Y + (3 - 3Y)$$

so

$$5X = 15$$

and

$$X = 3.$$

Substituting this into equation (1).

$$6 = 3 - 3Y$$

so

$$3 = -3Y$$

and

$$-1 = Y.$$

So the answer is $X = 3, Y = -1$.

As a check, equation (2) states $X = 4 + Y$ and 3 does equal $4 + (-1)$.

Just occasionally a pair of simultaneous equations has no solution. One such pair is

$$A = 3B + 2 \tag{1}$$
$$3A = 9B + 5 \tag{2}$$

If you multiply equation (1) by three and then subtract the equations you will find you have the result $0 = 1$, which is clearly not possible. Such a pair of equations has therefore no solution—there is no pair of numbers which can satisfy equation (1) *and* equation (2) simultaneously. Sometimes, on the other hand, a pair of simultaneous equations has no unique solution, for instance

$$A = 3B + 2 \tag{1}$$
$$3A = 9B + 6 \tag{2}$$

$A = 5, B = 1$ is one possible solution; $A = 2, B = 0$ is another; $A = -1, B = -1$ is another and so on. The problem is that the two equations are really the same: equation (1) is just equation (2) divided through by three. If, in modelling, you ever write down two apparently simultaneous equations which turn out to be insoluble or to have no unique solution you should go back and check that you have chosen your model correctly and then represented it by the correct equations.

SAQ 16

Suppose you were asked to solve

$$3P = 3 + Q \qquad (1)$$
$$7P = 2 + 4Q \qquad (2)$$

and that you planned to start by eliminating Q, which equation would you choose to multiply and by what number?

Perform this multiplication and hence solve the equations.

SAQ 17

Town X has a population of 52 000 increasing at an average of 5000 a year and Town Y has a population of 60 000 increasing at an average of 3000 a year. Use the model of SAQ 2 for each town and hence write down for each town an equation relating the population after some number of years to the number of years.

Find after how many years the populations are equal.

SAQ 18

A red car joins a motorway at an intersection and travels at a steady speed of ninety kilometres per hour. Half an hour later a blue car joins the motorway at the same intersection and travels at a steady 110 kilometres per hour.

Devise a model for each car to relate the distance it has travelled to the time elapsed since the *red* car joined the motorway, stating your modelling suppositions, and hence find how long after the red car joins the motorway it is overtaken by the blue car, and how far it has travelled on your model.

3.1 Solving all-letter simultaneous equations

In all the simultaneous equations you have met so far, the only items represented by symbols have been the two 'unknowns'; all the other items appearing have been numbers. This is rather analogous to the linear algebraic equations you solved in Unit 2 where the only item represented by a symbol was again the 'unknown'; all the other items were numbers. However, in Section 2 of this unit you learned to manipulate all-letter algebraic equations, and similarly in this section you will learn to manipulate all-letter simultaneous equations.

This can be useful because the solution is then in terms of parameters instead of numbers and can be used in many situations simply by substituting in the numerical values for the parameters which are appropriate to those situations.

For example, in Unit 2 you met the concepts of supply and demand and by plotting graphs you were able to obtain an estimate of the equilibrium price of a commodity. For every different commodity whose equilibrium price is to be found, this process has to be repeated and it would be useful if some sort of general expression could be found.

Figures 4(a) and (b) are included to remind you of the direction in which demand and supply curves slope. You should remember that they were modelled in Unit 2 by straight lines and that a straight line will always have an equation associated with it. After you have tried the following self-assessment question, I shall go on to describe the equations which can be associated with linear models of supply and demand.

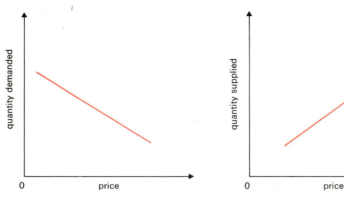

Figure 4

SAQ 19

SAQ 19

1   On the same axes plot graphs of the lines with equations

(a) $Q = 3 + 2P$

(b) $Q = 1 + 3P$

for values of P from 0 to 5, using a table of values for P and Q.

What is the gradient of line (a)?
What is the gradient of line (b)?
Where does line (a) cut the Q-axis?
Where does line (b) cut the Q-axis?

2   On the same axes plot graphs of the lines with equations

(a) $Q = 15 - 2P$

(b) $Q = 10 - P$

for values of P from 0 to 5.

What is the gradient of line (a)?
What is the gradient of line (b)?
Where does line (a) cut the Q-axis?
Where does line (b) cut the Q-axis?

Your work on SAQ 19 should have led you to appreciate that, in an equation of the form $Q = A + BP$, where A and B represent positive numbers, the graph of this equation cuts the Q-axis at A. This is called the *intercept* of the line on the Q-axis.

intercept

You should have also noticed that, in an equation of this form, B equals the gradient of the line.

Similarly, in an equation of the form

$Q = C - DP$

where C and D represent positive numbers, C is the intercept of the graph of this equation on the Q-axis and D is a measure of the steepness of the slope. In this case $-D$ equals the gradient. Also, the presence of the minus sign in front of D means that the line slopes *downwards* from left to right.

If you were going to model supply and demand by the equations

$Q = A + BP$

and

$Q = C - DP$

given A, B, C and D are all positive parameters, which equation would you choose for supply and which for demand?

The line representing the model of supply slopes up to the right, so $Q = A + BP$ is a suitable choice: the line representing the model of demand slopes down to the right, making $Q = C - DP$ a suitable choice.

The equation

$Q = A + BP$ (A and B positive)

is a general equation representing a linear model of supply and the equation

$$Q = C - DP \qquad (C \text{ and } D \text{ positive})$$

is a general equation representing a linear model of demand. In any specific case, values can be found for A and B or for C and D by reading off the intercept and gradient from the relevant graph.

To avoid confusion in what follows, I am going to denote the *demand* quantity by Q_D, so that Q_D stands for the total number of units demanded, and the *supply* quantity by Q_S, so that Q_S stands for the total number of units supplied.

Then

$$Q_S = A + BP$$
$$Q_D = C - DP$$

where P stands for the number of pence (say) in the price and A, B, C and D are all parameters, related to the graphs of supply and demand as described above.

> How do you think a general expression for the price at which all that is supplied will be bought, the equilibrium price, could be obtained from these general linear equations for supply and demand?

By finding and solving a pair of simultaneous equations.

At this particular price, the quantity which is demanded and the quantity which is supplied will be equal. Suppose that the number of units supplied or demanded at this price is $\bar{Q}$ and that the number of pence in this price is $\bar{P}$.

Then

$$\bar{Q} = A + B\bar{P} \tag{1}$$

and

$$\bar{Q} = C - D\bar{P} \tag{2}$$

Subtract equation (2) from equation (1).

$$\bar{Q} - \bar{Q} = A + B\bar{P} - (C - D\bar{P})$$
$$0 = A + B\bar{P} - C + D\bar{P}$$
$$0 = A - C + B\bar{P} + D\bar{P}$$
$$0 = A - C + \bar{P}(B + D)$$

So far I have followed exactly the same procedures as I would have done if I had been using numbers instead of A, B, C and D. I can make $\bar{P}$ the subject of the equation I have obtained by performing allowable operations.

Add C to both sides.

$$C = A + \bar{P}(B + D)$$

Subtract A from both sides.

$$C - A = \bar{P}(B + D)$$

Divide both sides by $(B + D)$—remember that B and D are both positive numbers, so $(B + D)$ is not zero.

$$\frac{C - A}{B + D} = \bar{P}$$

This gives the value for the price at which all that is supplied will be bought.

$$\bar{P} = \frac{C - A}{B + D}$$

SAQ 20

SAQ 20

For one commodity

$$Q_S = 10\,000\,000 + 500\,000P$$

and
$$Q_D = 20\,000\,000 - 300\,000P$$

Without solving these equations, use the result just derived to find a value for the equilibrium price.

SAQ 20 shows that the formula for the equilibrium price can be used for specific cases by substituting in the relevant values for A, B, C and D.

Once $\bar{P}$ is known, $\bar{Q}$ can be found. You know that
$$\bar{Q} = A + B\bar{P}$$

Therefore
$$\bar{Q} = A + B\left(\frac{C - A}{B + D}\right)$$

Putting the whole of the right-hand side over the common denominator $(B + D)$ gives
$$\bar{Q} = \frac{A(B + D) + B(C - A)}{B + D}$$

and multiplying out the brackets, remembering that whatever is outside the bracket multiplies *each* term in the bracket, gives
$$\bar{Q} = \frac{AB + AD + BC - AB}{B + D}$$

Therefore
$$\bar{Q} = \frac{AD + BC}{B + D}.$$

These general expressions for $\bar{P}$ and $\bar{Q}$ can be very useful for finding what happens to the equilibrium price when a tax is applied. If a flat-rate excise tax of T pence, paid by the buyer, is applied then you might think that the new equilibrium price would be $\bar{P} + T$, but this is not so. Since the *supplier* is still receiving $\bar{P}$ pence for his commodity he will still be supplying $\bar{Q}$ units. On the other hand, the *buyer* is paying $\bar{P} + T$ pence, so he will demand fewer units than $\bar{Q}$ (remember that as price increases demand decreases). There will therefore be an excess supply over demand, something that should tend to cause the price to drop, so the new equilibrium price will not be as high as $\bar{P} + T$.

It is possible to find the new equilibrium price as follows.

Suppose the new equilibrium price is $\bar{P}_1$ (read as 'P-bar one') and at this price $\bar{Q}_1$ ('Q-bar one') units are exchanged.

Using $Q_D = C - DP$ for demand will give
$$\bar{Q}_1 = C - D\bar{P}_1 \tag{3}$$
Using $Q_S = A + BP$ for supply will give
$$\bar{Q}_1 = A + B(\bar{P}_1 - T) \tag{4}$$

because the suppliers are receiving only the price the buyers pay *less the tax* when they supply the commodity.

Equation (4) can be rewritten as
$$\bar{Q}_1 = A - BT + B\bar{P}_1 \tag{5}$$

Equations (3) and (5) are very similar to equations (2) and (1), except that A in equation (1) has been replaced by $A - BT$ in equation (5). I can deduce therefore, without actually going through the process of solving the simultaneous equations (3) and (5), that
$$\bar{P}_1 = \frac{C - (A - BT)}{B + D}$$

$$= \frac{C - A + BT}{B + D}$$

$$= \frac{C - A}{B + D} + \frac{BT}{B + D}$$

I know that

$$\bar{P} = \frac{C - A}{B + D}$$

Therefore

$$\bar{P}_1 = \bar{P} + \frac{BT}{B + D}$$

This result shows that the price *has* gone up, but not by as much as T pence, because $B/(B + D)$ is less than 1 when D is a positive number.

SAQ 21

SAQ 21

Solve the simultaneous equations

$$Y = P + QX$$
$$Y = R + SX$$

given that $(Q - S)$ is not equal to zero.

3.2 Summary

In this section of the unit you have learned how to obtain pairs of equations that describe a model of a situation in which two related events occur and how to solve this pair of simultaneous equations, thus finding the point of coincidence of the two events.

Some simultaneous equations can be solved either by adding one of the two equations to the other or by subtracting one equation from the other in such a way that one 'unknown' is eliminated. In general, it is necessary first to multiply each term in one or both equations by a suitable number or parameter in order that one of the 'unknowns' may be eliminated when the equations are added or subtracted.

4 EQUATIONS INVOLVING DIMENSIONS

I mentioned in Unit 2 that it is possible to write down algebraic equations where the symbols represent both a number and a unit—that is, where the symbols represent *dimensioned quantities.*

All the equations you have met so far in the course have been of the type where a symbol represents a number. You may have noticed that all the equations have used capital letters for the symbols. This is because the course team have decided that in this course we will, in general, use capital letters to stand for numbers, and reserve small letters to stand for dimensioned quantities. In this way, we hope to help you to spot readily which kind of equation you are dealing with, since in this course some will be of one type and some of the other.

The first all-letter algebraic equation you met in Unit 2 was

$$C = PS$$

where C represents the number of pence seeding a lawn costs; P represents the number of pence cost per kilogram of seed; and S represents the number of kilograms needed.

In this equation, C, P and S represent numbers, but I could have defined the parameters differently so they were all dimensioned quantities. I could have chosen to use:

> the cost of seeding the lawn;
> the cost per unit mass of seed;
> the mass of seed needed.

'Cost' is a dimensioned quantity. Specific instances of 'cost' are 3 pence, 5 pounds, 20 dollars, 6 francs, and so on. In each of these instances, there is a number multiplied by a unit.

Similarly, 'mass' is a dimensioned quantity. Specific instances of 'mass' are 20 kg, 5 g, 100 lb, and so on. In each instance, there is a number multiplied by a unit.

'Cost per unit mass' is also a dimensioned quantity. Specific instances are 25 pence per lb, 4 francs per kg, and so on. In these cases the units are 'pence per lb' or 'francs per kg'. Once again there is a number multiplied by a unit.

Since my parameters are now dimensioned quantities, I need to use small letters to represent them. I shall let:

> c represent the cost;
> p represent the cost per unit mass;
> s represent the mass needed.

Then the equation relating these parameters is

$$c = ps$$

This equation turns out to be more generally useful than

$$C = PS$$

because C, P and S were all defined in terms of a specific set of units—costs *had* to be in 'pence' and masses in 'kilograms'. On the other hand, c, p and s are not defined in any specific set of units. The equation

$$c = ps$$

could be used by a Frenchman, who would want to use units of francs and kilograms, or by an American, who would want to use units of dollars and ounces. Each user of the equation

$$c = ps$$

must use a consistent set of units (as explained in Unit 2), but the equation itself does not confine the users to one particular set of units.

This explains why equations (like $c = ps$) involving dimensioned quantities are so useful to scientists and technologists—they are more 'general' than equations involving only numbers (like $C = PS$) and leave the user with the freedom to choose his set of units in any specific application.

There is a second reason why such equations are useful. Any dimensioned equation must be dimensionally balanced: if a modeller writes down an equation to represent some model he has made and then finds on checking that the equation is not dimensionally balanced, he can be certain that he has made an error. Thus he has an extra test of the validity of his model when he works with dimensioned equations.

To see what I mean by an equation being 'dimensionally balanced', I shall give a couple of examples.

In the equation $c = ps$,

c the cost, has dimensions of cost.

p, the cost per unit mass, has dimensions of cost/mass.

s, the mass, has dimensions of mass.

Thus the left-hand side of the equation $c = ps$ has dimensions of cost, while the right-hand side has dimensions of

$$\frac{\text{cost}}{\text{mass}} \times \text{mass}$$

But

$$\frac{\text{cost}}{\text{mass}} \times \text{mass}$$

just reduces to cost, since the mass can be divided out, and so the right-hand side of the equation $c = ps$ also has dimensions of cost. The equation is dimensionally balanced.

The second example is an equation relating to a model of the motion of some object which is approaching a fixed point at a steady speed

$$d = d_0 - vt$$

where d is the distance the object is from the point; d_0 is the distance it was from the point to start with; v is the speed; and t is time (i.e. how long it has been moving).

In an equation like this, it is not only necessary to check that the dimensions of the left-hand side are the same as those on the right, but also that each of the two terms on the right has the same dimensions. To try to subtract terms with different dimensions from each other is rather like trying to subtract two oranges from three apples—it is nonsense!

Both d and d_0 are distances and so they both have dimensions of length.

The speed v has the dimensions of length/time. This can be deduced from the fact that units for speed are kilometres per hour or metres per second.

The time t has dimensions of time.

Thus the left-hand side of the equation

$$d = d_0 - vt$$

has dimensions of length, and so has the first term on the right-hand side. The second term on the right has dimensions of

$$\frac{\text{length}}{\text{time}} \times \text{time}$$

which is also length, therefore the equation is dimensionally balanced.

Should an equation which a modeller has written down turn out to be dimensionally unbalanced when he checks it, the dimensional unbalance in the equation may help him to spot what it is that he has omitted. For instance, if the imbalance can be corrected by including a quantity with the dimensions of length/time, then since length/time are the dimensions of speed, he may well find that speed is the factor he neglected to feed into his equation.

Incidentally, you are now in a position to appreciate why the axis of a graph is labelled in the form distance/km or time/s, where s is the symbol for second(s), a unit of time. Since distance is a dimensioned quantity it represents a number multiplied by a unit. Only a number can be plotted on a graph, so the dimensioned quantity is divided by the unit (which is what distance/km means—remember the oblique stroke represents division) to leave just a number. Similarly, time is a dimensioned quantity, but time/s is just a number.*

Study comment

If you have found this section rather difficult, do not worry. The subject will become easier to you as you progress through the course and you see how useful and relevant it can be. Try SAQ 22, but do not spend too long on it: read the answer if you feel you are getting stuck and then go on to Section 5, which is a complete change of topic.

SAQ 22

SAQ 22

An equation used earlier in the unit is

$$P = N + AY$$

which was used to find the population of a town some number of years in the future. In its dimensioned form, this equation is

$$p = n + ay$$

where p is the population in the future; n is the population now; a is the population added per unit time; and y is the time until the population reaches p.

Check that this equation is dimensionally balanced.

(Hint: population can be thought of as having dimensions called 'population'.)

There is an alternative convention where dimensioned quantities may be plotted on graphs. You will meet this convention later in the course.

5 EXPONENTS

5.1 Compound interest

Suppose a man had £1000 to invest and left it untouched so that the interest was added to the capital at the end of each year. How much capital would he have each year if the rate of interest is 7%?

At the end of the first year his interest would be

$$1000 \times \frac{7}{100} \text{ pounds}$$

which would be added to his capital, making a capital for the second year of

$$1000 + 1000 \times \frac{7}{100} \text{ pounds}$$

which, taking out the common factor of 1000, is

$$1000\left(1 + \frac{7}{100}\right) \text{ pounds}$$

This can be rewritten as 1000×1.07, since

$$1 + \frac{7}{100} = 1.07.$$

At the end of the second year his interest would be

$$1000 \times 1.07 \times \frac{7}{100} \text{ pounds}$$

and his capital for the third year would be his capital during the second year plus this interest.

$$\left(1000 \times 1.07\right) + \left(1000 \times 1.07 \times \frac{7}{100}\right) \text{ pounds}$$

The two terms in this expression have a common factor of 1000×1.07, and so the expression can be rewritten as

$$1000 \times 1.07\left(1 + \frac{7}{100}\right) \text{ pounds}$$

which is

$$1000 \times 1.07 \times 1.07 \text{ pounds.}$$

Similarly, his capital during the fourth year would be

$$1000 \times 1.07 \times 1.07 \times 1.07 \text{ pounds.}$$

Clearly, 1.07 is going to multiply itself more and more times as the capital for each succeeding year is calculated, and it would be useful to have a shorthand way, a notation, to indicate that a number is multiplied by itself two, three, four, five...times. This can be done by using *exponents*.

1.07×1.07 is written $(1.07)^2$
$1.07 \times 1.07 \times 1.07$ is written $(1.07)^3$
$1.07 \times 1.07 \times 1.07 \times 1.07$ is written $(1.07)^4$

and so on.

How would $1.07 \times 1.07 \times 1.07 \times 1.07 \times 1.07 \times 1.07$ be written using an exponent?

$(1.07)^6$.

In the expression $(1.07)^4$ the number 1.07 is called the *base* and the number 4 the *exponent*.

What are the base and the exponent in $(1.07)^8$?

The base is again 1.07, and the exponent is 8.

Unfortunately, there is no standard terminology for the notation I have just introduced. The term 'base' is almost universally used, but the term 'exponent' is sometimes replaced by the term *index* or the term *power*, so that in $(1.07)^3$, the 3 could be referred to as an index or a power instead of an exponent, and the terms are generally used quite interchangeably.

Similarly, there is no standard way of referring to these expressions verbally, except in the case of the exponents 2 and 3.

$(1.07)^2$ is referred to as '1.07 squared'

$(1.07)^3$ is referred to as '1.07 cubed'

but $(1.07)^4$ can be referred to as '1.07 to the fourth power' or '1.07 to the fourth' or '1.07 to the power four'. Use whichever one you find easiest to say!

How can $(1.07)^6$ be referred to?

'1.07 to the sixth power', or '1.07 to the sixth' or '1.07 to the power six'.

Returning to the example of compound interest I introduced at the beginning of this section, the capital during the fourth year can be written $1000 \times (1.07)^3$ pounds.

How could the capital during the fifth year be written?

$1000 \times (1.07)^4$ pounds

My aim in introducing such an example has not been to teach you how to work out the capital the investor would have each year, but to show you that using exponents gives a more compact way of writing a long string of numbers like

$$1.07 \times 1.07 \times 1.07 \times 1.07 \ldots$$

I shall therefore discuss exponents in general, rather than in relation to an example, in the next sections. The results derived in these sections are summarized for ease of reference in Section 5.7.

5.2   Multiplication of numbers containing exponents

In Section 5.1, I used the base of 1.07 in all the numbers I expressed in *exponential form* because of the example I was using. Of course 1.07 is not the only base; any number, whether it is positive or negative, an integer* or a non-integer can be used as a base, with the exception of zero which is not used in certain circumstances which I shall mention later on.

For instance

$$2^4 = 2 \times 2 \times 2 \times 2 = 16$$

and

$$5^3 = 5 \times 5 \times 5 = 125.$$

---

* An integer is a whole number: 2, 5, 30, 151, −2, −6, are all integers.

Pair each of the numbers in the list A–G with its equivalent exponential expression in the list a–g.

A	36	a	5^4
B	81	b	6^2
C	10 000	c	10^4
D	32	d	$(\frac{1}{3})^3$
E	625	e	2^5
F	$\frac{1}{4}$	f	3^4
G	$\frac{1}{27}$	g	$(\frac{1}{2})^2$

If 4^3 is multiplied by 4^2, this is

$$4^3 \times 4^2 = (4 \times 4 \times 4) \times (4 \times 4)$$

which is

$$4 \times 4 \times 4 \times 4 \times 4 = 4^5$$

Therefore

$$4^3 \times 4^2 = 4^5 = 4^{3+2}$$

In *multiplying* 4^3 by 4^2 the exponents have been *added* together, so, *provided the bases are the same*, it is possible to carry out the multiplication of two (or more) numbers in exponential form by adding the exponents.

Suppose I multiply 3 by 3^4. This is $(3) \times (3 \times 3 \times 3 \times 3)$ which is 3^5.

If I want to generalize the rule that in carrying out the multiplication I add the exponents, then clearly 3 must be equivalent to 3^1. In general, any number to the power of one is equal to just the number itself, so $4^1 = 4$, $10^1 = 10$, and so on.

(a)   Write down values for the following, leaving your answer in the form involving an exponent.

   (i) $4^3 \times 4^2$

   (ii) $5^6 \times 5^7$

   (iii) $10^2 \times 10^4$

   (iv) $2^3 \times 3^2$

(b)   What are the numerical values of

   (i) 5^1

   (ii) $6^1 \times 7^1$?

5.3   Division of numbers containing exponents

If 2^5 is divided by 2^2, this is

$$\frac{2^5}{2^2} = \frac{2 \times 2 \times 2 \times 2 \times 2}{2 \times 2}$$

which is

$$2 \times 2 \times 2 = 2^3$$

Therefore

$$2^5 \div 2^2 = 2^3 = 2^{5-2}$$

In dividing 2^5 by 2^2 the exponents have been *subtracted*: the exponent of the number being *divided by* is subtracted from the exponent of the number that is being *divided into*.

As a second example

$$\frac{6^4}{6^3} = \frac{6 \times 6 \times 6 \times 6}{6 \times 6 \times 6}$$

which is 6.

Therefore

$$6^4 \div 6^3 = 6^{4-3} = 6$$

(remembering that 6 and 6^1 are equivalent).

So *provided the bases are the same* it is possible to carry out the division of two numbers in exponential form by subtracting the exponents.

Suppose I divide 4^5 by 4^7. If I use the rule I have just quoted, $4^5 \div 4^7 = 4^{-2}$ and 4^{-2} seems to be meaningless because of the negative exponent, but $4^5 \div 4^7$ can also be written as

$$\frac{4 \times 4 \times 4 \times 4 \times 4}{4 \times 4 \times 4 \times 4 \times 4 \times 4 \times 4} = \frac{1}{4 \times 4}$$

So I can generalize the rule if I let

$$\frac{1}{4 \times 4} = 4^{-2}$$

(4^{-2} is read as 'four to the minus two'.) This is how the meaning of a negative exponent is defined. For instance

$$5^{-3} = \frac{1}{5 \times 5 \times 5}$$

$$2^{-1} = \frac{1}{2}$$

$$3^{-4} = \frac{1}{3 \times 3 \times 3 \times 3}$$

and so on.

Zero is not used as a base with a negative exponent since $1/0$ is not defined.

SAQ 25 SAQ 25

(a) Write down values for the following, leaving your answer in a form involving an exponent.

    (i) $2^5 \div 2^2$ (iii) $5^6 \div 5^9$

    (ii) $3^6 \div 3^3$ (iv) $2^3 \div 3^2$

(b) What are the numerical values of

    (i) 4^{-1} (ii) 5^{-2}?

5.4   Defining a zero exponent

If I use the rules I have just described in the preceding two sections to multiply 5^3 by 5^{-3}, then I obtain a value of 5^0. An exponent of zero does not appear to have any meaning, but $5^3 \times 5^{-3}$ is also

$$(5 \times 5 \times 5) \times \frac{1}{5 \times 5 \times 5} = 1$$

The rules I have just introduced will still hold, even in this case, provided I define 5^0 to be equal to 1. In general, any base to the power zero is defined as being equal to 1 in order that the rules may always hold. Therefore

$$10^0 = 1 \qquad 3^0 = 1 \qquad 7^0 = 1 \text{ and so on.}$$

Once again, zero is not used as a base with a zero exponent.

5.5 Exponent form

Exponent form is a way of writing numbers that can make their manipulation easier. It is particularly convenient when very large or very small numbers are being manipulated, or when a slide rule is being used. It is for the latter purpose that I am introducing it here, as you will be using the slide rule in Section 6.

In exponent form, any positive number, no matter how large or small, is expressed as a number between 1 and 10 (to be precise, $1 \le$ number < 10, where $<$ means 'is less than') multiplied by a power of 10.

For example

$$430\,000 = 43\,000 \times 10$$
$$= 4300 \times 10^2$$
$$= 430 \times 10^3$$
$$= 43 \times 10^4$$
$$= 4.3 \times 10^5.$$

The exponent form of 430 000 is therefore 4.3×10^5. Similarly, the exponent form of 12 750 is 1.275×10^4.

When dealing with numbers less than 1, the exponent of 10 will be negative, but the procedure is the same. For example

$$0.0029 = 0.029 \times \frac{1}{10} = 0.029 \times 10^{-1}$$
$$= 0.29 \times \frac{1}{10} \times \frac{1}{10} = 0.29 \times 10^{-2}$$
$$= 2.9 \times \frac{1}{10} \times \frac{1}{10} \times \frac{1}{10} = 2.9 \times 10^{-3}$$

and the exponent form of 0.0029 is therefore 2.9×10^{-3}. Similarly, the exponent from of 0.0005 is 5×10^{-4}.

Unfortunately, the term 'exponent form' is by no means universally used. It is sometimes referred to as 'standard form' or as 'scientific notation' (this term is frequently used on electronic calculators) or as 'floating point form' (this term is more often used in connection with digital computers). This course will use the term *exponent form*.

The definition of exponent form as being a number in the range $1 \le$ number < 10 multiplied by a power of 10 is also not universal. Sometimes the number is chosen to be in the range $0.1 \le$ number < 1, sometimes it is chosen to be in the range $1 \le$ number < 1000. In this course, however, the course team will use, and ask you to use, the convention used in this section.

SAQ 26

(a) Write the following in exponent form.

(i) 2 390 000

(ii) 123

(iii) 0.012

(iv) 0.000 01

(v) 73.95

(b) Write out the following as ordinary numbers

(i) 3.2×10^5

(ii) 5×10^{-2}

(iii) 6.7×10^3

(iv) 7.85×10^{-1}

(v) 1.29×10^{-3}

If 430 000 is to be multiplied by 0.0029, this in exponent form is

$$4.3 \times 10^5 \times 2.9 \times 10^{-3} = 4.3 \times 2.9 \times 10^2$$
$$= 12.47 \times 10^2$$
$$= 1.247 \times 10^3.$$

If 12 750 is to be divided by 0.0005, this in exponent form is

$$\frac{1.275 \times 10^4}{5 \times 10^{-4}} = \frac{1.275}{5} \times 10^{4-(-4)}$$
$$= 0.255 \times 10^8$$
$$= 2.55 \times 10^7.$$

From this you can see that exponent form can help you to 'keep track of the noughts' in a calculation. This is particularly important when the slide rule is being used, as you will see in Section 6.

SAQ 27

SAQ 27

Express the following numbers in exponent form and perform the calculation, giving your answers in exponent form.

(a) 2300×400

(b) $350 \div 0.005.$

5.6   The graph of $Y = 2^X$

In order to continue this work with exponents, I need to plot the graph of an equation of the form

$$Y = A^X.$$

Suppose, for example, the graph is to be of the equation

$$Y = 2^X.$$

Table 1 shows some points on this graph and I have plotted these points in Figure 5.

Table 1

value of X	value of Y
-2	$2^{-2} = 0.25$
-1	$2^{-1} = 0.5$
0	$2^0 = 1$
1	$2^1 = 2$
2	$2^2 = 4$
3	$2^3 = 8$
4	$2^4 = 16$
5	$2^5 = 32$
6	$2^6 = 64$

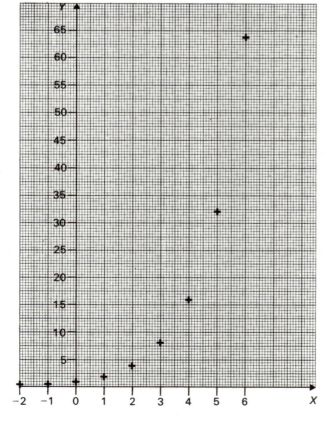

Figure 5

39

Nearly all of the graphs you have dealt with so far in this course have been straight lines, but clearly these points will not lie on a straight line. In fact, is it reasonable to join up the plotted points even with a curve? If I did so, as in Figure 6, has a point such as A whose coordinates are, as closely as can be read from the graph, (3.5, 11.3) any meaning? Remember that I am plotting $Y = 2^X$. If this point is to have any meaning then it must be that

$$11.3 = 2^{3.5}$$

to the accuracy to which the graph can be read.

Is this reasonable? $2^3 = 8$ and $2^4 = 16$, so $2^{3.5}$ should have a numerical value between 8 and 16, and 11.3 certainly is in this range. Further, if I apply the rules I have already quoted for multiplying numbers which incorporate an exponent I find that

$$2^{3.5} \times 2^{3.5} = 2^7 = 128$$

By multiplication

$$11.3 \times 11.3 = 127.69.$$

These two values are strikingly close, especially when we recall that 11.3 was the nearest value that could be read from the graph in Figure 6. It does look as though it is reasonable to say that $2^{3.5}$ equals 11.3, to the accuracy of the graph, and that a meaning can be given to non-integer exponents such as 3.5.

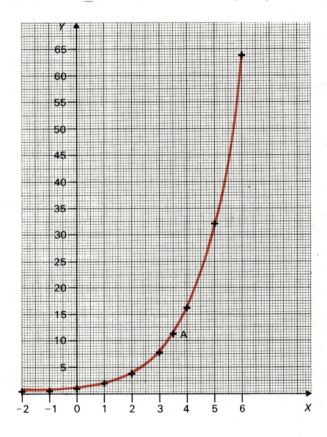

Figure 6

The graph could also be used to find (approximate) values for $2^{\frac{1}{2}}$, $2^{5/2}$, $2^{4.1}$ and so on.*

To find an approximate value for $2^{\frac{1}{2}}$, a line is drawn vertically up from the point $X = \frac{1}{2}$ to meet the curve, and then a line is drawn horizontally across. This line meets the Y axis at 1.4, so

$$2^{\frac{1}{2}} \simeq 1.4$$

where $\simeq$ means 'is approximately equal to'.

---

* $2^{\frac{1}{2}}$ is read as '2 to the half', $2^{5/2}$ is read as '2 to the five halves' or '2 to the five over two' and $2^{4.1}$ is read as '2 to the four point one'.

Similarly,

$$2^{5/2} \simeq 5.7$$

and

$$2^{4.1} \simeq 17.1$$

These are shown in Figure 7.

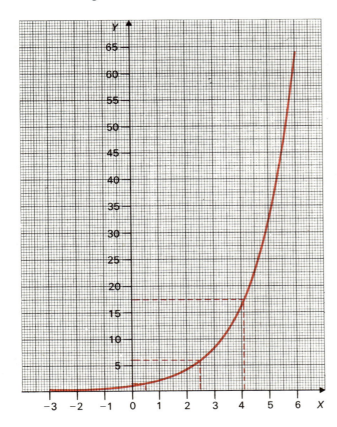

Figure 7

SAQ 28 SAQ 28

Use the graph of $Y = 2^X$ (Figure 6) to find approximate values for the following.

(a) $2^{3/2}$ (c) $2^{4.5}$

(b) $2^{2.2}$ (d) $2^{5.1}$

Suppose I apply the rule for multiplying numbers containing exponents to the case of $2^{\frac{1}{2}}$ multiplied by itself.

$$2^{\frac{1}{2}} \times 2^{\frac{1}{2}} = 2^{\frac{1}{2}+\frac{1}{2}} = 2^1 = 2.$$

You already know from Unit 1 that if a number multiplied by itself equals 2 then the number must be a square root of 2. I found from the graph that $2^{\frac{1}{2}}$ approximately equals 1.4, so if $2^{\frac{1}{2}}$ is indeed a square root of 2, then 1.4 should also be a square root of 2, and $(1.4)^2$ should equal 2. Of course, $(1.4)^2$ equals 1.96.

Does this mean that $2^{\frac{1}{2}}$ should not be a square root of 2? No! It simply means that 1.4 was only an approximation to $2^{\frac{1}{2}}$—the best value it was possible to obtain from the graph. A closer value is in fact 1.414.

Try evaluating $(1.414)^2$. Is its value 2?

$(1.414)^2$ is 1.999396, which is very nearly 2. 1.414 is clearly a closer approximation to a square root of 2 than was 1.4.

Clearly, the more digits that are correctly inserted after the decimal point in 1.414... the closer it will approximate to the exact value of the square root of 2. You may ask what this precise value for this square root of 2 is, but unfortunately, there is no way in which it can actually be written out! The square root of 2 is an example of an *irrational number*. Such numbers cannot be expressed as exact fractions and need an unending string of decimal digits to describe them precisely. There is not even a pattern in these digits which repeats itself (as there is for instance, in 0.030303...which is $\frac{1}{33}$ and is not irrational). By including more and more digits it is possible to come closer and closer to an exact value for an irrational number, but no finite string of digits can represent it exactly. Fortunately, it is seldom necessary to use a value for such a number which is any more precise than three significant figures.

irrational number

When 1.414...is denoted by $2^{\frac{1}{2}}$, then $2^{\frac{1}{2}}$ represents the *precise* value of the irrational number even though it can never be written out precisely as a decimal number.

Just as $2^{\frac{1}{2}}$ represents a square root of 2, $2^{5/2}$ represents a square root of 2^5 because

$$2^{5/2} \times 2^{5/2} = 2^5$$

What do you think $2^{3/2}$ represents?

A square root of 2^3, that is, a square root of 8.

If a number is represented as $2^{1/3}$ then it must be that
$$2^{1/3} \times 2^{1/3} \times 2^{1/3} = 2$$
and $2^{1/3}$ is called a *cube root* of 2. Similarly, $2^{1/4}$ is such that
$$2^{1/4} \times 2^{1/4} \times 2^{1/4} \times 2^{1/4} = 2$$
and $2^{1/4}$ is called a *fourth root* of 2.

cube root

What would $2^{1/5}$ and $2^{1/10}$ be called?

A fifth root of 2 and a tenth root of 2 respectively.

Similarly, $3^{\frac{1}{2}}$ is a square root of 3; $4^{5/2}$ is a square root of 4^5; $10^{1/3}$ is a cube root of 10; and so on. If a number is represented as $2^{5/6}$, then it must be that $2^{5/6} \times 2^{5/6} \times 2^{5/6} \times 2^{5/6} \times 2^{5/6} \times 2^{5/6} = 2^5$ so that $2^{5/6}$ is a sixth root of 2^5.

Figures 8(a) and (b) show two other curves which are similar to the graph of

$$Y = 2^X$$

which was drawn in Figures 6 and 7.

Figure 8(a) is

$$Y = 3^X$$

and Figure 8(b) is

$$Y = 10^X.$$

You can see that all these three curves have the same general shape. They rise very slowly for negative values of X, and more steeply as X becomes positive and larger. A curve of the form

$$Y = A^X$$

is called *an exponential curve* because X is called an exponent. There is one special member of this family of curves called *the* exponential curve and you will meet this curve in later units of the course when you will also investigate its properties. It is a curve that is often used in modelling.

exponential curve

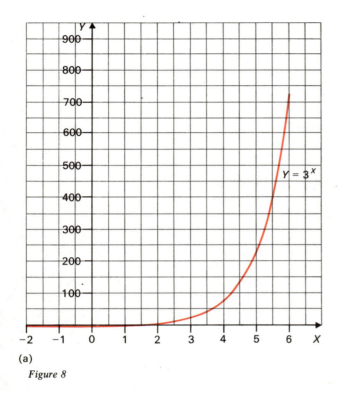

(a)

Figure 8

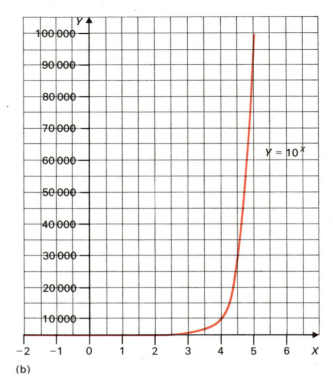

(b)

5.7 Summary

A number expressed in the form A^B is said to be expressed in an exponential form, A is called the base and B the exponent.

In multiplying two or more numbers expressed in exponential form and having the same base, the exponents are added.

In dividing two numbers expressed in exponential form and having the same base, the exponents are subtracted.

Whatever the value of A

$$A^1 = A.$$

For any value of A except $A = 0$

$$A^0 = 1$$

and

$$A^{-B} = \frac{1}{A^B}$$

$A^{P/Q}$ is called the Qth root of A^P, where P and Q are positive integers.

When a number is expressed in *exponent form* it is expressed in the form $N \times 10^M$, where $1 \leq N < 10$ and M is a positive or negative integer.

Exponential curves, that is graphs of equations of the form

$$Y = A^X$$

have the same general shape as the curves shown in Figures 7 and 8.

6 THE SLIDE RULE

The slide rule is a device which uses some of the properties of exponents you have met in Section 5 in order to make the processes of multiplication and division easier.

I want to approach the use of the slide rule by a simple analogy. Look at Figure 9 which shows two circular scales. Each scale shows numbers from 0 to 9.9 with subdivisions of 0.1.

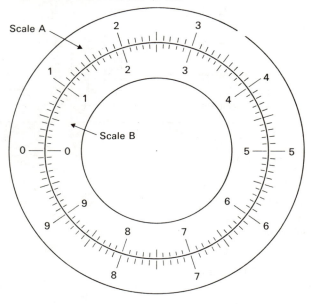

Figure 9

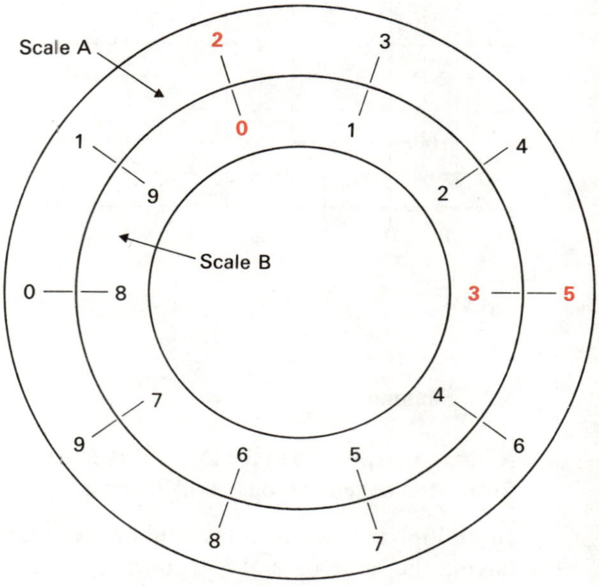

Figure 10

(subdivisions omitted for clarity)

Suppose that scale B can slide round inside scale A, as shown in Figure 10. Look at the situation in this figure, where the zero on the B scale is alongside the 2 on the A scale. You can see that the 3 on the B scale is alongside the 5 on the A scale.

Now $2 + 3 = 5$, so is the device adding?

Look at Figure 11(a), (b) and (c) and form your own conclusions.

I hope that you will have decided that the device is indeed adding. If it was used to add $X + Y$, the rule would be:

Slide the B scale round so that the zero on the B scale is alongside the X on the A scale. Look for Y on the B scale and the answer is alongside it on the A scale.

You may wonder what happens when the sum of the two numbers is greater than 10. What, for example, would happen if 6 was to be added to 7? Figure 12 shows this situation. Zero on the B scale is alongside 6 on the A scale. You can see that 7 on the B scale is alongside 3 on the A scale. Now $6 + 7 = 13$, so what has happened is that the device is giving the correct number of units, but is not giving the number of tens. This is a disadvantage, but not an insurmountable one, as you will see when you come to use your slide rule.

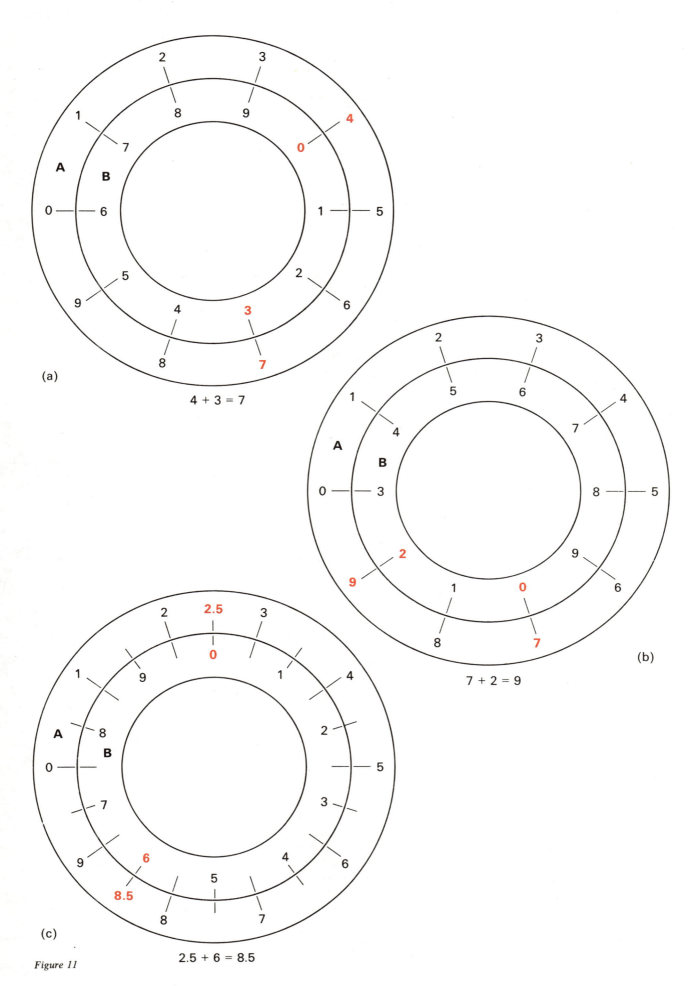

(a)

$4 + 3 = 7$

(b)

$7 + 2 = 9$

(c)

Figure 11

$2.5 + 6 = 8.5$

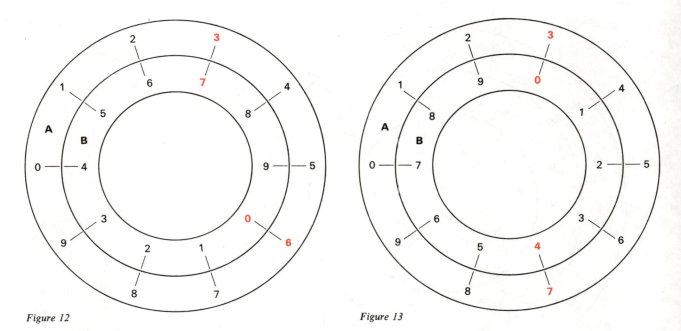

Figure 12

Figure 13

Now look at Figure 13 where I have shown a new arrangement which I can use for subtraction. To calculate 7 − 4, 4 on the B scale is put beside 7 on the A scale. The answer then appears on the A scale alongside the zero on the B scale.

Sketch the arrangement for calculating 9 − 3.2.

The sketch is shown in Figure 14.

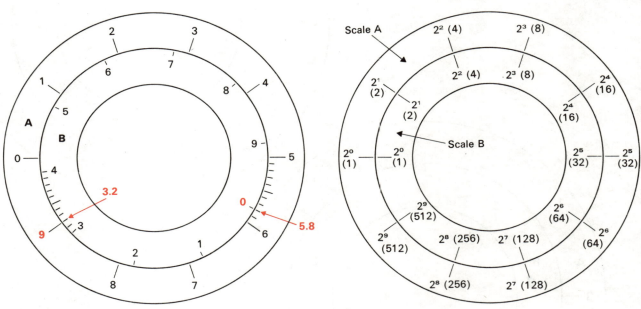

Figure 14

Figure 15

It is possible, therefore, to use sliding circular scales to add and subtract numbers correctly. They can also be used for multiplication and division, provided the process of multiplication and division is somehow transformed into addition and subtraction, respectively. You have already met one way of doing this in Section 5, where you learned that in *multiplying* two numbers which are expressed in an exponential form you *add* the exponents and in *dividing* two such numbers you *subtract* the two exponents. If two scales could be produced using numbers in exponential form, multiplication and division could be performed using addition and subtraction of the exponents. You will see that this idea is the basis of the slide rule.

Figure 15 shows a simple example of this, using a base of 2 for the exponential numbers. In Figure 16(a), you can see that this device is correctly multiplying $2^3 \times 2^4$ and in Figure 16(b) you can see that it is correctly dividing $2^8 \div 2^5$.

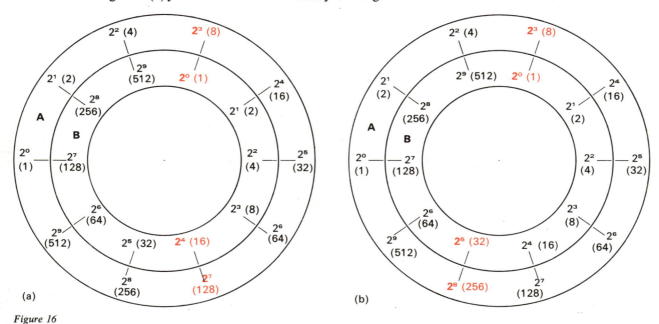

Figure 16

Most common slide rules, including the one you have been sent, use the base of 10. You will already have noticed that the circular scales on your slide rule are not linear (that is, they are not evenly spaced), and this arises because the scales are derived from exponents. In fact, they are derived by expressing each number in the form 10^X, as I shall show you.

Figure 17 shows the graph of $Y = 10^X$ for values of X between 0 and 1. You can see that the values of Y range from 1 to 10. Figure 18 is a blank straight scale 10 intervals long. This is going to be made into a scale suitable for a slide rule scale, using the graph in Figure 17.

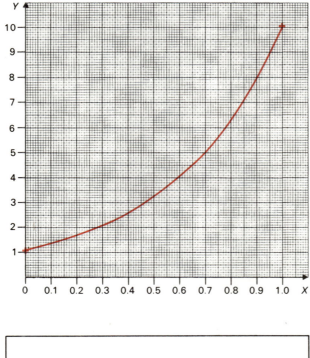

Figure 17

Figure 18

For example, $Y = 2$ corresponds approximately to $X = 0.30$, so the point 2 is put 0.3 of the way along the scale, or at 3.0. Similarly, $Y = 3$ corresponds to $X = 0.48$, so the point 3 is put 0.48 of the way along the scale, or at 4.8. This is shown in Figure 19.

Figure 19

SAQ 29

SAQ 29

Find the values of X corresponding to $Y = 1, 4, 5, 6, 7, 8, 9, 10$ by using Figure 17. Hence complete Figure 19 for all values of Y from 1 to 10 inclusive.

Compare the spacing of the numbers of the completed Figure 19 with the spacing of your slide rule. Is Figure 19 the basis for the scale of a straight slide rule (not the circular one you have)? By marking the points on the circumference of a circle, could you make Figure 19 the basis of the circular slide rule's scale?

You will have found that by scaling the numbers according to their representation in the form 10^x you have formed the basis of a slide rule. It is interesting to note that the slide rule is itself a kind of model—a physical model. The numbers on the scales of the slide rule you have been sent are represented in a convenient form for the purposes of carrying out multiplication and division. The representation is not suited to carrying out addition and subtraction because this is not one of the purposes of such a slide rule. By making a slide rule larger and larger it is possible to mark in more sub-divisions between the integers on the scales and thus to represent numbers more precisely along the scales.

6.1 Using the slide rule to multiply and divide

Now that you have seen how a slide rule is marked out to enable it to be used for multiplication and division, you are ready to start using your slide rule to carry out such manipulations.

If you are already familiar with how to use a slide rule to multiply and divide numbers then you should go straight to Section 6.2. Otherwise you should listen to Side 2 of Disc 1 which is a discussion programme on how to use the slide rule to multiply and divide. You should then study Sections 4 and 5 in your *Slide Rule Book* and try the accompanying exercises.

6.2 Summary

By using the properties of exponents, whereby adding the exponents of two numbers having the same base is equivalent to multiplying the numbers and subtracting the exponents is equivalent to dividing the numbers, it is possible to design a slide rule whose purpose is to enable multiplication or division to be carried out. The base used for the slide rule scales is 10.

SUMMARY OF THE UNIT

If the algebraic equation representing a model is to be used to reveal something about a situation which is not readily apparent then it is frequently necessary to *manipulate the equation* in some way. One useful manipulation which can be carried out is *changing the subject of the equation*. This is done by performing one or more *allowable operations* on the equation and is analogous to solving a linear equation with one 'unknown'.

Section 2.1

Brackets can be used in the manipulation of algebraic equations. A *common factor* in each of the terms on one side of an equation may be put outside the bracket and must multiply *each of the terms* inside the bracket. *Algebraic fractions* can be added and subtracted by using similar rules to those followed in adding and subtracting numerical fractions. A *common denominator* is found and the *numerator of each fraction is adjusted* to fit with its new denominator. The numerators can then be added or subtracted as appropriate.

Section 2.2

Sometimes a model can be represented by a *pair of linear equations*. Information about the point of coincidence of the situation being modelled can be found by solving the *simultaneous equations* derived from these two equations. Simultaneous equations can be solved by *adding* one equation to, or *subtracting* it from, the other equation in order to eliminate one of the 'unknowns'. Sometimes it may be necessary to *multiply one equation through* by some number or parameter before adding or subtracting the equations. In the case of supply and demand it is useful to have an expression for the equilibrium price in terms of four parameters—the *intercepts* of the supply and demand lines on the quantity axis and the *gradients* of the two lines. A specific value can be given to this equilibrium price by feeding in the appropriate values for these four parameters in any specific instance. Occasionally, a pair of equations may have *no solution*. In modelling, this usually points to an incorrect choice either of the model or of the equations representing the model.

Section 3

In an equation involving *dimensions* each symbol represents a *number multiplied by a unit*. In this course, such symbols will be *small letters*. Equations involving dimensioned quantities are usually more general than those involving only numbers. This is one reason for their usefulness. The other reason is that a check on the dimensional *balance* of the equation may reveal that some variable or parameter has been overlooked. A correct equation will always be dimensionally balanced.

Section 4

In the number 3^5, 3 is said to be the *base* and 5 the *exponent*. Alternative terms for exponent are *index* or *power*. This number 3^5 is referred to as 'three to the fifth power' or 'three to the fifth' or 'three to the power five'.

Section 5.1

In multiplying numbers in exponential form, *provided the bases are the same* the *exponents are added*. In dividing numbers in exponential form, *provided the bases are the same* the *exponents are subtracted*. For any value of A except zero, $A^0 = 1$ and $A^{-B} = 1/A^B$.

Sections 5.2, 5.3 and 5.4

In *exponent form* a number is expressed as a number between 1 and 10 ($1 \leq$ number < 10) multiplied by a power of 10.

Section 5.5

The graph of $Y = A^X$ has the general shape shown in Figures 6 and 8. These graphs illustrate *exponential curves*. Such graphs can be used to give numerical values of A^X where X is a non-integer. $A^{1/Q}$ is said to be the Qth root of A and $A^{P/Q}$ is said to be the Qth root of A^P. Such roots may be *irrational numbers*, as is for example the numerical value of $2^{\frac{1}{2}}$.

Section 5.6

The *slide rule* is a device whose purpose is to aid in the multiplication or division of numbers and the scales are chosen for this purpose. The scales are derived by expressing each number in the form 10^X and are *non-linear*. Linear scales would only be used if the slide rule was designed to add and subtract. Details on its use are given on *Disc 1* and in the *Slide Rule Book*.

Section 6

ANSWERS TO SELF-ASSESSMENT QUESTIONS

SAQ 1

(a) $2A - 15 = -1$
$2A = 14$
$A = 7$

(b) $3Z + 8 = 23$
$3Z = 15$
$Z = 5$

(c) $3(B + 1) = 5$
$B + 1 = 5/3$
$B = 2/3$

(d) $\dfrac{2X - 3}{5} = 5$
$2X - 3 = 25$
$2X = 28$
$X = 14$

(e) $2P + 2 = 5P - 4$
$2P + 6 = 5P$ or $2 = 3P - 4$
$6 = 3P$
$2 = P$

SAQ 2

(a) The number of inhabitants at present.
The number of inhabitants added per year.
The number of years from the present.
The number of inhabitants some years from the present.

(b) number of inhabitants some years from the present
= number of inhabitants at present
+ (number of inhabitants added per year
× number of years from the present)

(c) I shall let:
P stand for the number of inhabitants some years from the present.
N stand for the number of inhabitants at present.
A stand for the number of inhabitants added per year.
Y stand for the number of years from the present.

The algebraic equation, therefore, is $P = N + AY$.

(If you have chosen different letters check that your equation gives the same relationship between the parameters and variables.)

(d) Given that $N = 22\,000$ and $A = 200$:

Four years in the future
$P = 22\,000 + 200 \times 4$
$= 22\,000 + 800$
$= 22\,800.$

Five years in the future
$P = 22\,000 + 200 \times 5$
$= 22\,000 + 1000$
$= 23\,000.$

Eight years in the future
$P = 22\,000 + 200 \times 8$
$= 22\,000 + 1600$
$= 23\,600.$

Two years in the past
$P = 22\,000 + 200 \times (-2)$
$= 22\,000 - 200 \times 2$
$= 22\,000 - 400$
$= 21\,600.$

Three years in the past
$P = 22\,000 + 200 \times (-3)$
$= 22\,000 - 200 \times 3$
$= 22\,000 - 600$
$= 21\,400.$

Notice that negative values of time refer to time before the population was 22 000.

(e) The question states that *roughly* the same number of people is added each year, but in the model A is taken to be the same each year, so the model cannot give exact figures.

The model also supposes that the population goes on increasing as it has been doing, which may not be the case.

In addition to this, it may be that the population is not added at a steady rate throughout the year, but the model takes no account of this. In the model, if 200 are added per year, 100 are supposed to be added each half-year, fifty each quarter year, and so on.

SAQ 3

In the equation
$P = N + AY$
$P = 70\,000$
$N = 59\,000$
$A = 1200.$

Therefore

$70\,000 = 59\,000 + 1200\,Y$

Subtract 59 000 from both sides.

$11\,000 = 1200\,Y$

Divide both sides by 1200.

$\dfrac{11\,0\cancel{0}\cancel{0}}{12\cancel{0}\cancel{0}} = Y$

$Y = 9\tfrac{1}{6}.$

The population will be 70 000 about nine years from now.

SAQ 4

(a) Water resources will serve a town of up to 70 000 inhabitants, so

$Y = \dfrac{70\,000 - 65\,000}{1200}$

$= \dfrac{50\cancel{0}\cancel{0}}{12\cancel{0}\cancel{0}}$

$= 4\tfrac{1}{6}.$

Adjustments should be made to the water supply no later than about four years in the future, and preferably earlier. (Notice that a modelling supposition which has crept in is that usage per head will stay steady.)

(b) The Town Hall can deal with 21 000 households, which is equivalent to a population of 73 500, so

$Y = \dfrac{73\,500 - 65\,000}{1200}$

$= \dfrac{85\cancel{0}\cancel{0}}{12\cancel{0}\cancel{0}}$

$= 7\tfrac{1}{12}.$

The staff should be increased no later than about seven years in the future.

Notice that in both these cases the model gives an estimate of the number of years the present service will last, but I have given the answer in terms of 'no later than'. This is a case where an equality is being used in the model because, as I said in Unit 2, equalities are usually easier to deal with than inequalities. The answer is then being interpreted in a 'commonsense' way.

SAQ 5

(a) $D = 125T$

Divide both sides of the equation by 125.

$\dfrac{D}{125} = T.$

(b) $L = 110 + 5F$

Subtract 110 from both sides of the equation.

$L - 110 = 5F$

Divide both sides of the equation by 5.

$\dfrac{L - 110}{5} = F.$

(c) $V = U + AT$

Subtract AT from both sides of the equation.

$V - AT = U.$

(d) $V = U + AT$

$U + AT$ was formed by multiplying T by A and then adding on U, so the first step is to subtract U.

$V - U = AT$

Divide both sides of the equations by A (A is not zero).

$\dfrac{V - U}{A} = T.$

(e) $P = N + AY$

Subtract AY from both sides.

$P - AY = N.$

SAQ 6

(a) He needs to make P (the number of pounds each litre of paint costs) the subject of the equation.

(b) $C = \dfrac{PNA}{L}$

Divide both sides of the equation by N (N is not zero).

$\dfrac{C}{N} = \dfrac{PA}{L}$

Divide both sides of the equation by A (A is not zero).

$\dfrac{C}{NA} = \dfrac{P}{L}$

Multiply both sides of the equation by L.

$\dfrac{CL}{NA} = P$

Therefore

$P = \dfrac{CL}{NA}$

(c) Given that: $C = 9$, $L = 16$, $N = 2$ and $A = 48$

$P = \dfrac{^3\cancel{9} \times \cancel{16}}{2 \times \cancel{48}_{\,1}}$

$= \dfrac{3}{2}$

He can spend up to £1.50 per litre on paint.

SAQ 7

$C = \dfrac{PM}{K}$

Multiply both sides by K.

$CK = PM$

Divide both sides by C (C is not zero).

$K = \dfrac{PM}{C}$

SAQ 8

(a) $\dfrac{1}{X} = 4$

Multiply both sides by X.

$1 = 4X$

Divide both sides by 4.

$\dfrac{1}{4} = X$

(b) $\dfrac{5}{X} = 7$

Multiply both sides by X.

$5 = 7X$

Divide both sides by 7.

$\dfrac{5}{7} = X$

(c) $\dfrac{9}{2X} = \dfrac{1}{6}$

Multiply both sides by X.

$\dfrac{9}{2} = \dfrac{X}{6}$

Multiply both sides by 6.

$\dfrac{6 \times 9}{2} = X$

$27 = X$

(d) $\dfrac{3}{5X} = 1$

Multiply both sides by X.

$\dfrac{3}{5} = X$

SAQ 9

In

$$J = \dfrac{PNA}{L} + \dfrac{DNA}{M}$$

the common factor is NA, and

$$J = NA\left(\dfrac{P}{L} + \dfrac{D}{M}\right)$$

SAQ 10

(a) $S = \dfrac{UT}{2} + \dfrac{VT}{2}$

The common factor is $T/2$, and

$$S = \dfrac{T}{2}(U + V)$$

(b) $S = UT + \tfrac{1}{2}AT^2$

The common factor is T, and

$$S = T(U + \tfrac{1}{2}AT)$$

(c) $Y = 3X^2 - 2X$

The common factor is X, and

$$Y = X(3X - 2)$$

SAQ 11

(a) $Y = 3X^2 - X$
(b) $A = 10B - 14B^2$
(c) $D = 75T - 50$
(d) $I = MV - MU.$

SAQ 12

(a) $S = \dfrac{(U + V)T}{2}$

Multiply both sides of the equation by 2.

$2S = (U + V)T$

Divide both sides of the equation by $(U + V)$.

$\dfrac{2S}{U + V} = T$

(b) $S = \dfrac{(U + V)T}{2}$

Multiply both sides of the equation by 2.

$2S = (U + V)T$

Divide both sides of the equation by T.

$\dfrac{2S}{T} = U + V$

Subtract U from both sides of the equation.

$\dfrac{2S}{T} - U = V$

(Notice that here I could not treat $(U + V)$ as a single entity throughout the calculation, as I needed to pick out one of the terms in the expression $(U + V)$ and make it the subject of the equation.)

SAQ 13

(a) $\dfrac{1}{X} + \dfrac{2}{Y}$

A common denominator is XY, so

$\dfrac{1}{X} + \dfrac{2}{Y} = \dfrac{Y}{XY} + \dfrac{2X}{XY}$

$\qquad = \dfrac{Y + 2X}{XY}$

(b) $\dfrac{1}{A} - \dfrac{B}{C}$

A common denominator is AC, so

$\dfrac{1}{A} - \dfrac{B}{C} = \dfrac{C}{AC} - \dfrac{AB}{AC}$

$\qquad = \dfrac{C - AB}{AC}$

(c) $\dfrac{P}{QR} - \dfrac{S}{R}$

A common denominator is QR, so

$\dfrac{P}{QR} - \dfrac{S}{R} = \dfrac{P}{QR} - \dfrac{QS}{QR}$

$\qquad = \dfrac{P - QS}{QR}$

SAQ 14

$P = L\left(\dfrac{C}{NA} - \dfrac{D}{M}\right)$

A common denominator of the terms in the bracket is NAM, so

$P = L\left(\dfrac{CM}{NAM} - \dfrac{NAD}{NAM}\right)$

$\quad = L\left(\dfrac{CM - NAD}{NAM}\right)$

$\quad = \dfrac{L}{NAM}(CM - NAD)$

SAQ 15

The answers given illustrate one way of finding each solution.

(a) $X = 2Y - 3$ (1)

  $3X = 2Y + 5$ (2)

Subtract X from the left-hand side of equation (2) and $2Y - 3$ from the right-hand side.

$3X - X = 2Y + 5 - (2Y - 3)$

If I subtract -3 from 5 the result is 8, so

$2X = 8$

and

$X = 4.$

Substituting this into equation (1)

$4 = 2Y - 3$

so

$\dfrac{7}{2} = Y.$

The solution is $X = 4$, $Y = 7/2$.

As a check, equation (2) states $3X = 2Y + 5$ and 3×4 does equal $(2 \times 7/2) + 5$.

(b) $2A = 5B + 2$ (1)

  $2A = 3B - 2$ (2)

Subtract $2A$ from the left-hand side of equation (2) and $5B + 2$ from the right-hand side.

$2A - 2A = 3B - 2 - (5B + 2)$

$0 = -2B - 4$

$2B = -4$

$B = -2$

Substituting this into equation (1)

$2A = -10 + 2$

$2A = -8$

$A = -4$

The solution is $A = -4$, $B = -2$.

As a check, equation (2) states $2A = 3B - 2$ and $2 \times (-4)$ does equal $[3 \times (-2)] - 2$.

SAQ 16

  $3P = 3 + Q$ (1)

  $7P = 2 + 4Q$ (2)

Equation (1) should be multiplied by four, giving

$12P = 12 + 4Q$

Subtract $12P$ from the left-hand side of equation (2) and $12 + 4Q$ from the right-hand side.

$7P - 12P = 2 + 4Q - (12 + 4Q)$

$-5P = -10$

so

$P = 2$

Substituting this into equation (1).

$6 = 3 + Q$

so

$Q = 3$

The solution is $P = 2$, $Q = 3$.

As a check, equation (2) states $7P = 2 + 4Q$ and 7×2 does equal $2 + (4 \times 3)$.

SAQ 17

Let P_X stand for the number of inhabitants in Town X after some years and let Y stand for the number of years; then, using the model of SAQ 2

$$P_X = 52\,000 + 5000\,Y$$

Similarly, if P_Y stands for the number of inhabitants in Town Y after some years,

$$P_Y = 60\,000 + 3000\,Y$$

If the populations are equal after $\bar{Y}$ years, and if the populations are then $\bar{P}$

For Town X

$$\bar{P} = 52\,000 + 5000\,\bar{Y}$$

For Town Y

$$\bar{P} = 60\,000 + 3000\,\bar{Y}$$

These form a pair of simultaneous equations.

$$\bar{P} = 52\,000 + 5000\,\bar{Y} \qquad (1)$$
$$\bar{P} = 60\,000 + 3000\,\bar{Y} \qquad (2)$$

Subtracting $\bar{P}$ from the left-hand side of equation (2) and $52\,000 + 5000\,\bar{Y}$ from the right-hand side

$$\bar{P} - \bar{P} = 60\,000 + 3000\,\bar{Y} - (52\,000 + 5000\,\bar{Y})$$
$$0 = 8000 - 2000\,\bar{Y}$$
$$2000\,\bar{Y} = 8000$$
$$\bar{Y} = 4$$

So the populations are equal after four years.

SAQ 18

The model which you will almost certainly have used is one supposing both cars travel at a constant speed and that they reach that speed as soon as they join the motorway.

Let T stand for the number of hours that have elapsed since the red car joined the motorway and D_R represent the number of kilometres it has travelled from the motorway intersection.

The red car travels at ninety kilometres per hour, so: when one hour has elapsed the car has travelled 90×1 kilometres; when two hours have elapsed the car has travelled 90×2 kilometres; and when T hours have elapsed the red car has travelled $90 \times T$ kilometres. Therefore

$$D_R = 90 \times T$$

The restriction $T \geq 0$ should be included.

$$D_R = 90 \times T \qquad (T \geq 0)$$

Let D_B represent the number of kilometres the blue car has travelled from the motorway intersection. The blue car travels at 110 kilometres per hour, so one hour after the red car joined the motorway is half an hour after the blue car joined and this blue car has travelled $110 \times \frac{1}{2}$ kilometres. Two hours after the red car joined the motorway is $1\frac{1}{2}$ hours after the blue car joined and it has travelled $110 \times 1\frac{1}{2}$ kilometres. T hours after the red car joined is $(T - \frac{1}{2})$ hours after the blue car joined and it has travelled $110(T - \frac{1}{2})$ kilometres. Therefore

$$D_B = 110(T - \tfrac{1}{2})$$

The restriction $T \geq \frac{1}{2}$ should be included, since the blue car joins the motorway half an hour after the red car joins it:

$$D_B = 110(T - \tfrac{1}{2}) \qquad (T \geq \tfrac{1}{2}).$$

When the blue car overtakes the red car, they have both travelled the same number of kilometres from the intersection, say $\bar{D}$ kilometres, and it is, say, $\bar{T}$ hours since the red car joined. Then, for the red car

$$\bar{D} = 90\,\bar{T}$$

and for the blue car

$$\bar{D} = 110(\bar{T} - \tfrac{1}{2})$$

so

$$\bar{D} = 110\,\bar{T} - 55.$$

These form a pair of simultaneous equations.

$$\bar{D} = 90\,T \qquad (1)$$
$$\bar{D} = 110\,\bar{T} - 55 \qquad (2)$$

Subtract equation (1) from equation (2).

$$0 = 20\,\bar{T} - 55$$
$$20\,\bar{T} = 55$$
$$\bar{T} = 2.75$$

The overtaking takes place 2.75 hours after the red car joined the motorway.

Substitute $\bar{T} = 2.75$ into equation (1).

$$\bar{D} = 90 \times 2.75$$
$$= 247.5$$

The cars have travelled 247.5 kilometres along the motorway.

These values are estimates and are only as accurate as the model.

SAQ 19

1 You were asked to plot the graphs for values of P from 0 to 5 and therefore the tables of values for equation (a) and equation (b) are as given in Tables 2 and 3.

Table 2   Table of values for (a) $Q = 3 + 2P$

P	$2P$	$3 + 2P = Q$
0	0	3
1	2	5
2	4	7
3	6	9
4	8	11
5	10	13

Table 3   Table of values for (b) $Q = 1 + 3P$

P	$3P$	$1 + 3P = Q$
0	0	1
1	3	4
2	6	7
3	9	10
4	12	13
5	15	16

Figure 20(a) shows the graphs.

The gradient of line (a) is 2.

The gradient of line (b) is 3.

Line (a) cuts the Q-axis at $Q = 3$.

Line (b) cuts the Q-axis at $Q = 1$.

2 Figure 20(b) shows the graphs.

The gradient of line (a) is -2.

The gradient of line (b) is -1.

(The negative values arise because the lines slope *down* to the right.)

Line (a) cuts the Q-axis at $Q = 15$.

Line (b) cuts the Q-axis at $Q = 10$.

SAQ 20

The price at which all supplies will be bought can be found by substituting appropriate values for A, B, C and D into

$$\bar{P} = \frac{C - A}{B + D}$$

Comparing

$$Q_S = 10\,000\,000 + 500\,000\,P$$

with

$$Q_S = A + BP$$

you can see that

$$A = 10\,000\,000$$
$$B = 500\,000$$

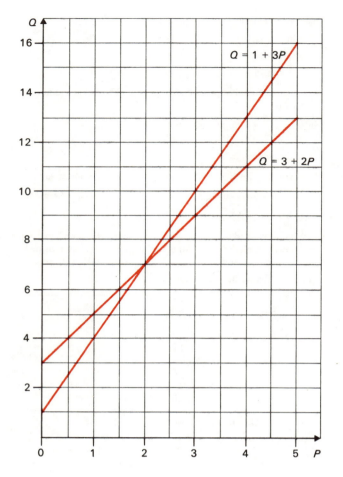

(a)

Figure 20

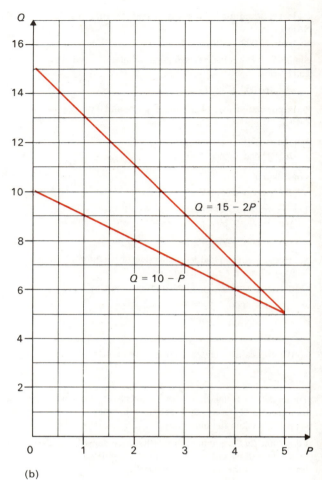

(b)

Similarly, comparing

$$Q_D = 20\,000\,000 - 300\,000\,P$$

with

$$Q_D = C - DP$$

you can see that

$$C = 20\,000\,000$$
$$D = 300\,000$$

Therefore

$$\bar{P} = \frac{20\,000\,000 - 10\,000\,000}{500\,000 + 300\,000}$$

$$= \frac{10\,000\,000}{800\,000}$$

$$= \frac{100}{8} = 12.5$$

The equilibrium price is $12\frac{1}{2}$ pence.

SAQ 21

$$Y = P + QX \tag{1}$$
$$Y = R + SX \tag{2}$$

Since Y and X are the symbols which appear in both equations, they must represent the two 'unknowns'.

Subtract equation (2) from equation (1).

$$Y - Y = P + QX - (R + SX)$$

Therefore

$$0 = P + QX - R - SX$$
$$0 = P - R + X(Q - S)$$

Add R to both sides.

$$R = P + X(Q - S)$$

Subtract P from both sides.

$$R - P = X(Q - S)$$

Divide both sides by $(Q - S)$, which is not equal to zero.

$$\frac{R - P}{Q - S} = X$$

You may have chosen to subtract equation (1) from equation (2), in which case you will have obtained

$$X = \frac{P - R}{S - Q}$$

Since

$$\frac{R - P}{Q - S} = \frac{P - R}{S - Q}$$

these results are identical.

Substitute

$$X = \frac{R - P}{Q - S}$$

into equation (1).

$$Y = P + Q\left(\frac{R - P}{Q - S}\right)$$

$$= \frac{P(Q - S) + Q(R - P)}{Q - S}$$

$$= \frac{PQ - PS + QR - QP}{Q - S}$$

54

$$= \frac{QR - PS}{Q - S}$$

Similarly, you may have obtained

$$Y = \frac{PS - QR}{S - Q}$$

which is correct since

$$\frac{PS - QR}{S - Q} = \frac{QR - PS}{Q - S}$$

SAQ 22

The left-hand side of the equation, p, has dimensions of population. The first term on the right-hand side, n, also has dimension of population. In the second term on the right-hand side of the equation:

a has dimensions of $\dfrac{\text{population}}{\text{time}}$

and

y has dimensions of time.

Therefore, ay has dimensions of

$$\frac{\text{population}}{\text{time}} \times \text{time}$$

which reduces to the dimensions of population.

Therefore the equation is dimensionally balanced.

SAQ 23

A b
B f
C c
D e
E a
F g
G d

SAQ 24

(a) (i) 4^5

    (ii) 5^{13}

    (iii) 10^6

    (iv) $2^3 \times 3^2$. Exponents can be added only if the bases are the same.

(b) (i)   5

    (ii) 42 (because $6^1 \times 7^1 = 6 \times 7$, which is 42).

SAQ 25

(a) (i) 2^3

    (ii) 3^3

    (iii) 5^{-3}

    (iv) $2^3 \div 3^2$. Exponents can be divided only if the bases are the same.

(b) (i) $\dfrac{1}{4} = 0.25$

    (ii) $\dfrac{1}{5 \times 5} = \dfrac{1}{25} = 0.04.$

SAQ 26

(a) (i) 2.39×10^6

    (ii) 1.23×10^2

    (iii) 1.2×10^{-2}

    (iv) 1×10^{-5} or just 10^{-5}

    (v) 7.395×10 or, strictly, 7.395×10^1.

(b) (i)   320 000

    (ii) 0.05

    (iii) 6700

    (iv) 0.785

    (v) 0.001 29

SAQ 27

(a) $2300 \times 400 = 2.3 \times 10^3 \times 4 \times 10^2 = 9.2 \times 10^5.$

(b) $350 \div 0.005 = \dfrac{3.5 \times 10^2}{5 \times 10^{-3}}$

                  $= 0.7 \times 10^5$

                  $= 7 \times 10^4.$

SAQ 28

(a) 2.8

(b) 4.6

(c) 23

(d) 34.

Do not worry if you have not been able to read Figure 6 to quite this accuracy, so long as your answers are similar to these.

SAQ 29

The values are as follows:

Y	X
1	0
4	0.60
5	0.70
6	0.78
7	0.85
8	0.90
9	0.95
10	1.0

Once again, do not worry if you did not obtain *exactly* these values.

The completed Figure 19 is shown in Figure 21. Your comparison of this scale with your circular slide rule should show that the spacing follows a similar pattern with the scale becoming more cramped in the region of the larger numbers. A complete comparison is not possible, of course, because your slide rule is circular, but it does appear that Figure 21 is the basis of a straight slide rule and that if the points had been marked on the circumference of a circle you would instead have had the basis of a circular slide rule.

Figure 21

Modelling by Mathematics